The University of Alberta Press

DICK DEKKER

with photographs by

EDGAR T. JONES

Wildlife at

Beaverhills Lake,

Alberta

Prairie Water

This revised edition first published by
The University of Alberta Press
141 Athabasca Hall
Edmonton, Alberta, Canada T6G 2E8

5 4 3 2 1

Canadian Cataloguing in Publication Data

Dekker, Dick, 1933-
Prairie water

Includes bibliographical references.
ISBN 0-88864-308-X

1. Birds—Alberta—Beaverhill Lake. 2. Ecology—
Alberta—Beaverhill Lake. Natural history—Alberta—
Beaverhill Lake. I. Jones, Edgar T. II. Title.
QL685.5.A86D44 1998   598.097123'3   C98-910551-2

∞ Printed on acid-free paper.
Printed and bound in Canada by Quality Color Press,
Edmonton, Alberta.

The University of Alberta Press acknowledges the financial
support of the Government of Canada through the Book
Publishing Industry Development Program for its
publishing activities. The Press also gratefully acknowledges
the support received for its program from the Canada
Council for the Arts and the Alberta Foundation for the Arts.

COMMITTED TO THE DEVELOPMENT OF CULTURE AND THE ARTS

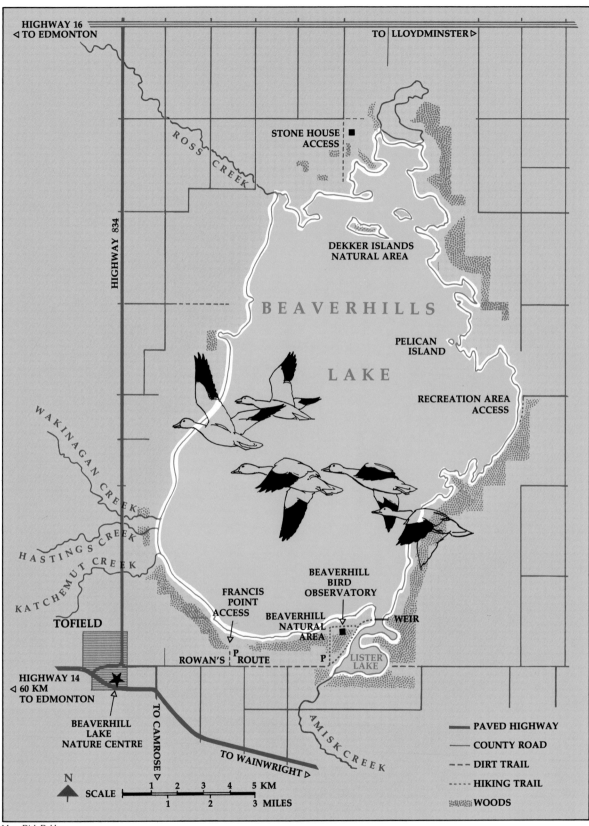

HIGHWAY 16
◁ TO EDMONTON

TO LLOYDMINSTER ▷

ROSS CREEK

HIGHWAY 834

STONE HOUSE
ACCESS ■

DEKKER ISLANDS
NATURAL AREA

B E A V E R H I L L S

PELICAN
ISLAND

L A K E

RECREATION AREA
ACCESS

WAKINAGAN CREEK

HASTINGS CREEK

KATCHEMUT CREEK

BEAVERHILL
BIRD
OBSERVATORY

FRANCIS
POINT
ACCESS

BEAVERHILL
NATURAL
AREA

WEIR

TOFIELD

ROWAN'S P ROUTE

P
LISTER
LAKE

HIGHWAY 14
◁ 60 KM
TO EDMONTON

BEAVERHILL
LAKE
NATURE CENTRE

TO CAMROSE ▷

TO WAINWRIGHT ▷

AMISK CREEK

N
SCALE

1   2   3   4   5 KM
1   2   3 MILES

PAVED HIGHWAY
COUNTY ROAD
DIRT TRAIL
HIKING TRAIL
WOODS

Map: Dick Dekker

CONTENTS

With its Canadian breeding range restricted to the larger cattail marshes in interior B.C. and the prairie provinces, the yellow-headed blackbird is a typical representative of the lake's western avifauna. As if to make up for the poor melodious quality of its song, the bird's spectacular gold-and-black plumage gives it an almost exotic look. (See also cover photo).

FOREWORD

Beaverhills Lake, at Tofield, Alberta, is no longer a secret among those who seek out the natural treasures of our land. Its designation as a Ramsar site places it in the company of the most important wetland environments in the world and promises that local stewardship will ensure its natural integrity on behalf of the world community. As a premium bird-watching location, it is mentioned in both national and international bird-finding guides. And the south shore is replicated in museum galleries.

During several years of doing research on birdwatchers at Point Pelee National Park in Ontario, I was frequently surprised to meet overseas "birding ecotourists" who, having crossed the Atlantic or Pacific Oceans for the spring birding at Pelee, also included a visit to Beaverhills Lake as part of their Canadian itinerary.

While it lies within a commuter's drive of nearly a million people, very few of them have visited Beaverhills. It doesn't draw the usual water-based recreation groups, and it lacks the rugged majesty of the Rockies or the bizarre uniqueness of the Badlands. As an attraction for wildlife, its assets are obvious to the animal world, but to the human community the Beaverhills landscape requires a subtle eye and an informed mind. With the proper introduction, the magic of the place will bring you back time and again. This work is that introduction, and Dick Dekker will unfold to you the magic of Beaverhills as few others could.

He has roamed the shores of Beaverhills in the tradition of a Henry David Thoreau, yet with the eye of a modern ethologist. His observations are rich and astute, and his descriptions of behaviour and interactions are the product of not only a skilled observer but a patient one. Generally, his travels are solitary, moving like one of the wolves he has tracked so extensively in Jasper. The pages which follow are the product of these peregrinations, which by his estimate constitute perhaps 14,000 kilometres of foot travel over nearly three decades. Compressed to actual field days, they constitute four solid years of daily observations. This is twice the time Thoreau spent at Walden's Pond.

And Beaverhills is Dekker's Walden. He, like Thoreau, has lived on it and explored it with a passion. In the process he has given us, in *Prairie Water*, a collection of his observations and insights that will unfold to all of us the magic of this special place which, as he describes, can be found wherever "we set our sights, on the horizon or on the grass at our feet."

Dr. Jim Butler
Professor of Parks, Wildlife and Interpretation
University of Alberta

INTRODUCTION AND ACKNOWLEDGEMENTS

Although most visitors focus keenly on birds, the lake's underlying attraction is the scenery, the vast panorama of water and sky, and the uncluttered horizons. There are no cottages or private beaches to impede access, and in its rural setting, the lake casts a spell of peaceful solitude.

But despite its large size, it is just a shallow slough of muddy water, unsuitable for boating, swimming or fishing, and of no use to people at all, it seems, except to birdwatchers and duck hunters. The shoreline scenery is not spectacular. It varies little in a subtle merging of narrow beaches, rock-strewn points, and reed-choked bays. Cow pastures surround the lake, tree-less and open in the north and west; covered with scrubby poplar and willow in the south and east.

The lake takes its name from the Beaver Hills, a wooded ridge of higher land a few kilometres to the west, but it is usually called Beaverhill Lake, in the singular form, as if there were a prominent landmark rising from its shore that inspired the name. Perhaps, the misnomer was the consequence of a mistake made many years ago by a cartographer who dropped the "s" from the original name, Beaver Hills Lake. A century ago, it was also known as Beaver Lake, which some people continue to call it to this day. Nevertheless, the name that seems to make the most sense is the plural form, Beaverhills Lake, which will be used in this book, and which was also the preference of Robert Lister, who watched birds at the lake for over 60 years. His book "The Birds and Birders of Beaverhills Lake" is a treasure of anecdotes based on Lister's own observations and on the journals of his associate, Professor William Rowan, who shot and collected birds at the lake from 1920 to 1957.

In contrast to Lister's book, this publication aims to give a more focussed look at the lake's birds, as well as some perspectives on habitats, plants and mammals. However, this book does not give a complete description of the flora and fauna of the lake. Much standard information can be found in the popular field guides and in the handbooks, such as "The Birds of Alberta" by W. R. Salt and J. R. Salt, or "The Birds of Canada" by W. E. Godfrey. These chapters are intended to be readable and entertaining, with an emphasis on the behaviour of birds and hints on field identification of some difficult species that are much sought after by visiting birdwatchers.

Unless stated otherwise, the opinions expressed in this book, on issues ranging from the implications of habitat management to the duck-hunting strategies of the peregrine falcon, are those of the author, who spent more than 1400 days at the lake during all seasons of the year. Each one of these days included more than ten kilometres of walking along all sections of the shoreline.

Special thanks are due to local farmers who permitted the "odd duck" from the city to roam over their land, carrying only binoculars, not a gun as some feared. Prior to 1965, people on foot were considered suspect and unwanted, and their parked cars might be vandalized to discourage further visits. However, after the landowners became familiar with the phenomenon of birdwatching, their tolerance and generosity in allowing access onto their shoreline pastures were unfailing.

The author's field work on birds of prey at the lake from 1973 to 1984 was financially supported by the Alberta Fish and Wildlife Division, World Wildlife Fund Canada, and the Alberta Recreation, Parks and Wildlife Foundation. The latter institution also financed the writing of this manuscript, which was gathering dust on the author's desk until the the fall of 1990 when Mike Quinn, coordinator for Prairie For Tomorrow, obtained a grant towards the book's publication. Prairie For Tomorrow was a joint venture of World Wildlife Fund Canada and the Alberta Fish and Wildlife Division (Buck For Wildlife Program).

The following people provided pertinent information or reviewed the chapters on specific subjects; Ken Bailey (Ducks Unlimited Projects), Derek Johnson (Botany), Ken Lungle (Duck Lure Program), Stefan Jungkind (Songbirds, BBO), Wayne Roberts (Fish and Amphibians), Bruce Turner (Duck Hunting and Banding), Bill Wishart (Canada Geese), and Dale Wrubleski (Chironomids). Thanks are also due to the authors of publications listed in the Literature Section, as well as to those who reviewed or contributed records for the Checklists. Peter DeMulder read the manuscript and made helpful comments.

The sleek Western grebe is a common diving bird on prairie water. Expert birdwatchers may argue that this particular individual looks like a Clark's grebe, since the black cap does not extend below the (fiery red) eyes.

Chapter 1

FROM GLACIERS TO FARMS

During the billion-year history of our planet, the region that is now called central Alberta was subjected to extreme changes in climate and topography. As the continents heaved and settled, the interior of North America was alternately flooded by oceans and tropical seas, or buried under arctic icecaps.

Although primitive forms of life existed much earlier, land animals did not emerge until some 400 million years ago, as recorded by their fossilized remains buried in sediments of ancient marshes. *Homo sapiens* came to North America relatively recently, some ten to thirty thousand years ago, when the Pleistocene merged into the present geological epoch, the Holocene. It is believed that the ancestors of American natives crossed the Bering Sea via a land-bridge between Asia and Alaska, and that they gradually worked their way down the Pacific coast or through the Rocky Mountain valleys to the interior plains of Canada and beyond.

Archaeologists have unearthed evidence that stone-age people lived in central Alberta soon after the last continental ice sheet retreated, some eleven thousand years ago. The mile-high glacier, that had covered all of the province with the exception of western mountain tops and some southern hills, stagnated east of present-day Edmonton, where the melting ice deposited its debris of boulders and gravel in a hummocky moraine that is now called the Beaver Hills. Icebergs, large and small, that were covered over by the debris during the glacier's retreat, melted later, causing the overlying material to settle into bowl-like depressions that filled with water. This peculiar landscape of rolling gravel piles and potholes is termed ''knob and kettle'' topography.

It is doubtful that the early Albertans cared for or understood the region's geographic origins. They probably led a nomadic life and were preoccupied with the hunt for caribou and muskoxen in the tundra-like landscape that unfolded in the glacier's wake. There is evidence that the climate became quite mild and arid later, since the mosses and lichens of the tundra were succeeded by grasses and trees in a savannah-like habitat. A colder and wetter regime set in some three thousand years ago, when the Beaver Hills became covered with boreal forests of spruce, tamarack, birch and poplar. Caribou and muskox had long since moved north, following the ice cap, while bison, moose and elk became the dominant herbivores of the region, sustaining carnivores, such as wolves, as well as the human hunters.

The ''old people'' left little record of their existence; a few stone tools and weapons, and some crude red-ocre paintings on cliffs or the walls of caves. Their unrecorded culture ended in chaos some three centuries ago, when the human population of Canada entered a violent stage of expansion, that was followed by permanent changes on the land, deeply affecting its plants and animals. This revolution was sparked by the arrival on the Atlantic coast of a new race of immigrants, the intrepid Europeans, their mission the conquest and plunder of the ''new world''.

In late winter, before the return of waterfowl, the lake shore scenery is starkly beautiful in its own right when wind-sculpted drifts frame rotting ice, or meltwater pools reflect the sky. Photos: Dick Dekker

Shock waves reached the Alberta hinterland in the early 1700s, when eastern Indian tribes, armed with rifles, axes and knives, traded from the "white man", ventured west and south from their homelands in Ontario and along James Bay. The primitive Alberta natives were displaced and annihilated. Continuous inter-tribal warfare and contagious diseases such as smallpox killed over half of western Indians around 1780. The Crees became the new masters of the Beaver Hills. Their advance to the south was halted by the horse-mounted warriors of the Blackfoot confederacy, who dominated the prairies beyond the Battle River, where many a bloody clash took place.

The name Beaver Hills is translated from the Cree "Amisk Wachi". It is not clear whether this name is in reference to the Beaver tribe that originally lived here, or to the furry rodents that were common in the lakes, ponds and streams. However, it is certain the beaver were the region's most sought-after treasure. In 1795, when the fur-trading North-West Company reached central Alberta and built Fort Edmonton, the Indians called the establishment "Beaver Hills House".

The fort was supplied with meat and pemmican, the voyageur's staple, by Metis hunters who travelled to and beyond the Beaver Hills in search of bison. They worked in large parties to discourage attacks from bands of Indians. Despite the heavy kill from natives, Metis and whites, the bison remained common in the region until the 1870s, when a new wave of immigrants began arriving: the settlers. In the next century, they caused much waste and destruction. Bison were exterminated while all other herbivores and fur-bearers became scarce. Virgin sod was turned, old forests set on fire. Only the odd island in the larger lakes escaped the conflagration that filled the air with smoke during all seasons. Ironically, many settlers who permanently scarred the Beaver Hills, found the shallow soil unsuitable for agriculture. Today, much of the area has returned to forest, albeit predominantly poplar and aspen instead of the majestic conifers of a century ago.

However, the parklands and the "buffalo wool prairies" east of the Beaver Hills became a settler's paradise. The foot-deep black soil ranks among the most fertile in the province with a high content of organic matter and nitrogen.

In 1890, after the railroad had linked the Canadian east with the western provinces, the influx of settlers into central Alberta increased dramatically. The government opened a lands agency in Beaver Lake settlement, which was later named after its first full-time doctor, J. H. Tofield, who set up practice in the pioneer district in 1894. In the first decade of the 20th century, the town of Tofield experienced booming land sales. Coal mines were opened up nearby and remained in business until 1953. In their search for good water, a perennial concern in this region of salty ground water, Tofield drillers inadvertently struck natural gas in 1910. It caused a flurry of speculation and town council installed gas flares along main street. The gas petered out a few years later, but deeper wells drilled in the 1970s were more productive and gas industry activity is now conspicuous on the fields and pastures that surround the lake.

The lake itself has been viewed by settlers and present-day residents with mixed feelings. The changing water levels make the "big slough" a natural resource of dubious value. Desirable fish species were plentiful only in the first decades of this century. And waterfowl hunting is allowed just a few months of the year. However, the lake's true merits are beginning to be appreciated. In 1987, the town of Tofield opened its Nature Centre and officially recognized the economic possibilities of birdwatching. This renewed emphasis on the area's wildlife can only bring benefits for people and birds alike, and holds the promise that the lake will be preserved in its natural state for future generations.

Chapter 2

WATER LEVELS AND THE AQUATIC FOOD CHAIN

When you stand on the lake's south shore and look north across its 18 kilometre expanse, there is a stretch on the horizon where water touches sky. It's the closest thing to an ocean view in land-locked Alberta. The vastness of the lake is a pleasant discovery for a first-time visitor. But other aspects may make a less favourable impression, especially if the name "Beaverhills" had evoked images of precipitous vistas.

There are those who find the flat scenery boring, or who are disgusted by the cow dung littering the muddy shores. Even life-time patrons such as Robert Lister, who knew and loved the lake for over sixty years, did not fall under its spell at first sight. His book "The Birds and Birders of Beaverhills Lake" includes a hilarious account of a 1920 swimming expedition when he found the tepid water not deep enough to cover his knees, making it hard to escape from the torment of mosquitoes and horseflies. His first chapter begins thus: "The lake lay flat as glass under the heat of the afternoon sun. The haze shimmering over the water veiled the far shore, so that one looked out over what appeared to be a strangely quiet sea. Only where the wavelets lapped listlessly on the mud and sand was there any motion, and from where they rippled in the wrack came a strong stench of rotten weed. This was my first look at Beaverhills Lake, and I was not impressed.... If there were any waterfowl, and there surely must have been, I do not remember them. All I recall is the stench and the excruciating itch I suffered all the following day and night."

The stench referred to by Lister probably originated from algae blooms that can coat sections of shore with a substance resembling blue-green paint. Poisonous to cattle, the algae seem harmless to birds that wade or swim through it. Its obnoxious smell is pervasive far downwind from affected areas. The itch Robert suffered after his "swim" was caused by the larvae of a bird parasite common to most shallow lakes on the prairies, especially after warm spells when it multiplies rapidly. The organism penetrates human skin and causes so-called "swimmer's itch" that can last for days.

A characteristic of Beaverhills Lake that annoys surrounding land-users is the fluctuation of its water levels. The earliest recorded account of low water dates from 1865 when a traveller saw bison mired in the mud of the exposed lake bottom. By contrast, at the turn of the century, a series of extremely wet years boosted the water table by five metres, and the lake reached its largest size recorded in known history. It was then an integral link in the so-called Cooking Lake watershed that drained most of the Beaver Hills and included half a dozen other sizeable lakes: Miquelon, Larry, Oliver, Joseph, Ministic, and Hastings. Interconnected by creeks, the lakes discharged into the largest of them all, Beaverhills Lake, which in turn drained into the North Saskatchewan River via Beaver Creek, that was up to 40 metres wide and two metres deep in the early 1900s.

Like all of the Cooking Lake watershed, Beaverhills Lake began a drying trend in the

1920s, until Beaver Creek stopped running and the lake became only a collection basin without outlet. Incoming creeks (Ketchamoot, Maskawan, and Hastings on the west side; Ross on the northwest; and Amisk on the southeast) now flow only spasmodically during spring run-off and after heavy rains.

The deterioration of the watershed may be partly attributable to a drying trend in the climate, but man-made changes on the land no doubt accelerated the process and may have prevented recovery. The destruction of the original Beaver Hills forest adversely affected nature's water regulatory system. Instead of percolating through a mossy forest floor, snow melt and summer rains now cascade across cleared ground or are wasted by increased evaporation in surface pools and puddles. In 1926, the decline in lake levels was hastened by the digging of a ditch from Miquelon to Camrose to supplement the town's failing water supply. In the thirties and forties, all lakes in the area continued to shrink in size, and Beaverhills Lake almost dried up completely during the 1950s.

As will be explained later, some cyclic fluctuation of the lake level is a must for the maintenance of productive bird habitat, but farmers are understandably upset when their lands are alternately flooded or turned into mudflats that may impede or prevent access to water for cattle. Irritated by such problems in the past, land-users have publicly expressed their desire to prevent excessive lake fluctuations by the construction of outflow canals with flood control weirs and pumping stations. The first petition to that effect was presented to the government in 1917, after the lake had risen during two very wet years. However, before the engineers had finished a preliminary survey, nature had lowered the lake level to what it was before 1915.

In 1969, after Edo Nyland wrote his article ''This Dying Watershed'' that alerted the public to the deterioration of the chain of lakes in the Beaver Hills, many groups and individuals took a serious look at an exciting new idea: the restoration of the Cooking Lake watershed to

its former glory by bringing in water via a pipeline from the North-Saskatchewan River. The grand scheme ran aground on opposition from landowners, who did not want their shoreline properties flooded.

In 1974, following the melt of near-record winter snow, nature dealt many farmers around Beaverhills Lake a soggy blow when the lake rose about one metre. Predictably, the Lease Holders Association clamoured for swift action. The Department of the Environment responded by initiating a study to explore the feasibility of installing an outlet canal on the lake. The projected cost, to be borne by the public purse, proved to be economically unjustified and the scheme was not implemented. The farmers did not protest, because, in the meantime, nature had done the job of lowering the water to what it had been before.

On the surface, stabilization of the capricious lake levels seems to make sense, since it should prevent flood problems for farmers as well as for some ground-nesting bird species, namely gulls, pelicans and cormorants, that lay their eggs on islands in the lake. However, stabilization would not benefit species that depend on open shoreline and mudflats, either for nesting or feeding. These littoral habitats would gradually disappear as a result of the expansion of reed beds, a process that has been ongoing for a dozen years since 1977, a period when the lake has varied little in level. Inexorably, and unnoticed by all but a few regular visitors, bulrushes and cattails have sent out their creeping roots until the plants closed ranks along most of the west, north and east shores.

The invasion of reeds will only be halted when a major drought and dropping water levels leave the reeds high and dry, killing their roots. Drought has additional benefits for the ecosystem. It quickens the break-down of dead plant material, freeing the nutrients, and it gives the seeds of emergent vegetation a chance to dry out and germinate.

Periods of high water have a different and complementary set of benefits. It is a drastic and

Quite apart from its bird life, the lake's underlying attraction is the vast panorama of water and sky. Lake level fluctuations are essential for the creation of mudflat habitat and the natural control of reeds and bulrushes.

Photos: Dick Dekker

effective way of ridding the mudflats of invading grasses and sedges. Exceptionally high floods can even kill reeds, as well as willows and poplars farther up the shore. The dead and decaying plant material adds to the detritus on the lake bottom and serves as food for invertebrates and in turn for their predators, including fish and birds.

In summing up, water level fluctuations are a vital component of the food chain in a dynamic wetland system. Man-made schemes to stabilize the lake should not be implemented unless they include periodic and gradual draw-downs to imitate the natural trophic cycles. If the timing of the alternating floods and droughts can be manipulated with understanding as well as engineering skill, the habitat manager may even improve on nature's erratic and heavy-handed ways.

Perhaps more serious than the proposals for stabilization of the water table, was a suggestion to drain the entire lake and turn the area into a giant community pasture. This plan arose in the 1950s after the lake had shrunk far back, leaving vast mudflats that turned into a sea of grass the following summer. Haying was going on kilometres beyond the former shore.

Other ideas for development involved the mining of coal seams that underly the area, and the possible use of the lake's vast expanse as a shooting range for the airforce. Fortunately, neither of these drastic schemes got further than the proposal stage. Future suggestions for manipulation or destruction of the lake will no doubt come under close scrutiny from various government agencies, while private conservation groups would be much more vigilant and vocal in their opposition than they were formerly, when environmental concerns played a very minor role and when the lake was much less widely known and appreciated.

Recent developments that have enhanced the lake's status, and that can only help to ensure adequate protection, are the designation of two provincial Natural Areas that are to be left inviolate for the future. One

of the two is centered on the Pelican and Dekker Islands in the northeast of the lake, the other consists of a strip of land along the southeast shore on which the Bird Observatory is located. In 1982, the lake gained national recognition when it was declared a National Nature Viewpoint by the Canadian Nature Federation. In Wildlife Year 1987, when the two Natural Areas were dedicated, the lake was also designated as a Wetland of International Importance under the Ramsar Convention. Named after the location of its first meeting, the city of Ramsar in Iran, the international group intends to fight the accelerating loss of wetlands around the world. To date, more than half of the planet's marshes, estuaries, peatlands and bogs have been drained, destroyed or heavily polluted. Beaverhills Lake is Alberta's fourth Ramsar Site, after the Peace-Athabasca Delta, the Hay-Zama Lakes and the Alberta portion of the muskegs which are the nesting grounds for whooping cranes. In total, Canada has 28 Ramsar Sites with over 13 million hectares, more than half of the 22 million hectares listed by 45 countries that are signatories to the treaty as of 1987.

Ramsar designation is not an attempt by the international group to prevent all environmental change or management, and it does not prejudice the sovereign rights of the Province of Alberta, but it is a way of ensuring that the potential hazards of development are thoroughly examined through international concern and pressure.

On the surface, the future of Beaverhills Lake seems bright and secure. Under the surface, the picture is less clear, literally and figuratively. Since the lake loses water only through evaporation, its chemical content will concentrate over time. Creeks and run-off from surrounding farmland will continue to bring in traces of natural salts, fertilizers, herbicides and pesticides, that cannot be flushed out. Some of the pollutants are absorbed by reeds and rushes, that have the capacity to purify water, while excess nutrients will enrich the growth of aquatic plants and algae, rendering the water milky and turbid with suspended material. The dissolved soda salts, sodium carbonates and sulfates make the water somewhat alkaline, with a pH value of 8-9, which acts to neutralize the insidious perils of acid rain. Clear northern lakes, that are not alkaline, do not have such built-in protection against airborne pollution, and they accumulate acids until the water becomes virtually sterile, unable to support fish or fish-eating birds.

In Beaverhills Lake, fish have a survival problem for quite a different reason. The shallow waters freeze to the bottom each winter, killing all fish unless they can locate a deep spot with sufficient oxygen content. It is possible that there are a few holes, perhaps fed by springs that run all year, but most of the lake is so shallow that a canoeist can travel for miles without finding any water deeper than the blade of a paddle.

At the turn of the century, when the lake brimmed with clear water, whitefish and pike occurred in marketable quantities. Local people netted them by the wagon load, and one farmer, who had built a fish trap in Ketchamut Creek, used the daily catch to feed his pigs. He sold the meat on the Edmonton market until the public refused to buy anymore pork that tasted of fish.

As the lake level began dropping in the 1920s, winter-killed fish littered the shores at break-up time, and eventually whitefish and pike disappeared altogether. A few small species have managed to survive: brook stickleback and fathead minnow, none bigger than 6-10 cm. A spectacular increase in their numbers occurred in 1974 after the lake had risen about one metre. The following summer, millions of young fry schooled over inundated pastures and amongst drowned willows near the southeast dam. In spring, mature sticklebacks and minnows darted up creeks and meltwater ditches in search of suitable spawning habitat. But their hordes dwindled after 1977 when the lake waters went down again, and by 1987 all spawning runs were reduced to a minimum.

While the increasing scarcity of small fish was tied to the dropping lake levels and the bottoming out of the nutrient cycle in the late 1980s, man-made changes to the lake's inlets played a role too. For instance, in the fall of 1984 a water control weir was constructed in a ditch connecting a system of sloughs and dug-outs to the main lake. The following spring, thousands of sticklebacks and minnows strained against the dammed-up water that cascaded down a culvert. Unable to overcome the metre-high obstruction, the fish died en masse. Two years later, there were no fish at all in this ditch.

In nearby Ross Creek, fish migrations befell a similar fate after county crews dynamited and bulldozed an upstream series of beaverdams during the winter of 1984-1985. In April, when the ice went out, dead minnows and sticklebacks lined the snow-rimmed banks along the creek's outflow over half a kilometre. An average of 100-300 carcasses per metre brought a rough estimate of dead fish to 100,000. In addition, unknown numbers had washed down the current onto the still frozen bay of the lake, where gulls and crows, as well as three bald eagles, had a feast. Since 1987, fish migrations up Ross Creek are very minor.

The Ducks Unlimited weir that was built to maintain and create waterfowl breeding habitat in Lister Lake, seems to have been a mixed blessing for fish. During the first few years after the weir's closure, sticklebacks and minnows massed against the concrete barrier, vainly skittering up the rivulets that flowed overtop. Reverse movements took place during fall, but were also halted. Thousands upon thousands of little fishes swarmed in the shallows upstream, unable to enter the main lake. Eventually, they froze into the ice in a solid mat of bodies. After the mid 1980s, a few minnows could be seen on the downstream side of the weir, but larger schools occurred in Lister Lake, where the water is probably deep enough in spots for the winter survival of some fish.

The fluctuations in fish had a profound effect on fish-eating birds. During the mid

17

1970s, grebes, gulls and terns were exceptionally numerous; pelicans, cormorants and herons set up breeding colonies. A few years later, all of these species became less common or stopped nesting, and some changed their diet from fish to amphipods, commonly called sideswimmers or scud. These whitish, shrimp-like crustaceans are near the top of the aquatic food chain, which contains about thirty species of smaller crustaceans and more than a hundred species of insects, mollusks, worms, rotifers, water mites and hydra, as well as single-celled protozoans, bacteria and algae. All these lifeforms, in their weird and wonderful array of shapes and sizes, feed on each other and are part of the obscure but vital process of recycling nutrients and energy in the aquatic environment. Their periodic abundance ensures that the shallow lake is a great place ''for the birds''.

Photo: Dick Dekker

Photo courtesy: Ducks Unlimited

In the 1970s, the lake's southeast bay was closed off with a dam and weir to control and stabilize the water level in Lister Lake, which was enhanced with islands as potential breeding sites for waterfowl.

Chapter 3

THE BIRDING SEASONS

The first migratory birds return to central Alberta in late February, well ahead of the official end of winter. The harbinger of the coming season is not the celebrated goose or the hardy crow, but a tiny inconspicuous songbird, the horned lark. Chirping brightly, it flutters along gravel country roads, the only bare ground in a snow-covered landscape.

The progress of spring can make a great difference from year to year. Winter snow may be gone in early March, or remain well into April. Extensive snowcover does not deter geese, but their arrival coincides with a spell of mild weather that produces some open water in Amisk Creek or Lister Lake. Most early-arriving honkers are mated pairs, that compete with each other for suitable nesting platforms. Their earliest recorded date is the first of March, but they are usually a week or two later.

Ducks are not far behind. As they fly smartly over the wintry countryside, or splash down in a puddle, their colourful spring plumage is a feast to the eye. Mallards and pintails are tied for first place, well ahead of a dozen other species of dabblers and divers.

Transient geese, on their way to the arctic and subarctic, come in waves that hug the weather fronts, travelling at night and lifting off again in the evening when a tail wind clears the northern sky. Canadas arrive well before white-fronts and snows. On peak days, their numbers are in the tens of thousands. In late seasons, the big Canadas press on quickly, the mass of them passing through in a few days, but very large mixed flocks of small

northern Canadas, white-fronts and snows linger well into the second week of May.

The spectacular migrations of sandhill cranes are equally uneven and can easily be missed. The great majority of birds, in mile-long flocks, passes on a few days in late April or early May, while stragglers continue to soar over for two weeks.

Weather conditions are also of crucial importance to the movements and build-up of shorebirds. Before the lake is free of ice, sometime between mid-April and early May, yellowlegs, godwits, dowitchers and pectoral sandpipers congregate on flooded fields and wet meadows. During cold springs, the migrating flocks come and go, apparently finding little food. By contrast, if the weather is warm and there is a lot of meltwater on the land, shorebird numbers build up but may be dispersed over a wide area. A dry and warm spring, with little meltwater, favours the concentration of shorebirds along the lake, especially if the nutrient cycle is high.

Sandpipers are usually most numerous along muddy and open sections of the west shore, upwind of the predominant westerlies. The wave-swept strip of sand along the south shore attracts sanderlings, while dowitchers, yellowlegs and stilt sandpipers favour muddy bays. Black-bellied plover, knots and turnstones frequent the rock-strewn points and overgrazed pastures along the east and west side of the lake. Towards the middle of May, when the weather warms, the number and variety of shorebirds increase, often coinciding with a massive hatch of midges, a

Photos: Dick Dekker

In 1989, when these photos were taken, the southeast bay began to fall dry and the mudflats became overgrown. Snow melt and spring rains brought the lake level up again in 1997.

large species of chironomid, locally called lake-flies. During cool springs, when the major hatch of midges does not take place until the end of May, sandpipers and plovers linger well into June. When the lake's trophic cycle is low and chironomids are scarce, the mudflats and pastures are deserted before the end of May.

After the departure of transient shorebirds, only locally breeding species remain: killdeer, avocet, marbled godwit, willet and Wilson's phalarope. Their protest calls greet the visiting birdwatcher as long as their young are unable to fly.

On and over the water, many other locally nesting birds invite attention: family groups of Canada geese, bachelor parties of canvasback drakes, fussing ruddy ducks and quarreling coots. Colonies of eared grebes, yellow-headed blackbirds and Franklin's gulls are hidden among the reed beds. Ring-billed and California gulls nest on islands where they are safe from mammalian raiders. A few cormorants and white pelicans crowd together on a tiny islet shared with the gulls. Other large fish-eating species, such as the black-crowned night-heron and the bittern, conduct their family affairs discreetly in the cover of reeds.

While many of the local birds are still caring for their young, the first migrants from the arctic are on the way back. By the middle of July, least sandpipers appear on the mudflats, soon to be joined by a few Baird's and semi-palmated sandpipers, their identification a challenge for the watcher. Far out on the lake, red-necked phalaropes begin to gather, but they never reach the incredible numbers that occur in May.

By mid August, some corners of the lake are crowded with yellowlegs, dowitchers, stilt sandpipers and a few pectorals. Hudsonian godwits mingle with the last remaining marbled godwits or willets. An occasional whimbrel may be among them too, but less regularly than in spring. On the pastures, a few buff-breasted and upland sandpipers may be discovered, or the odd golden plover among the more common black-bellies. White-rumped sandpipers and dunlin are far from regular, but a few are seen each spring and fall.

In recent years, while the number of competent birders visiting the lake has increased, a growing list of extreme rarities has been sighted: surfbird, sharp-tailed sandpiper, ruff, western sandpiper, ancient murrelet, sabine's gull, little gull and long-tailed jaeger. The chance of spotting a new or unusual species enhances the birdwatcher's day at the lake.

Late summer and fall are good times for the raptor enthusiast. Red-tailed and Swainson's hawks soar over the fields or sit on posts. The plumage of newly-fledged juveniles is quite different from the adults, requiring careful identification. When food was plentyful during the past breeding season, harriers and short-eared owls produced many young. Families of kestrels fly from post to post, hunting dragonflies, and quarreling with merlins that also hunt the large insects. Peregrines and prairie falcons may turn up too, scattering flocks of sandpipers and blackbirds. Especially during late fall, there is a small chance of spotting a gyrfalcon.

Other bird hunters, such as the sharp-shinned hawk, pass by over the wooded east shore, soaring and circling, or darting among the willows. Cooper's hawks and goshawks occasionally hunt low among the reed beds, chasing shorebirds and ducks. A most unexpected hunter of sandpipers is the parasitic jaeger, that usually migrates along the oceans, but has been recorded often at the lake. Mid September is perhaps the best time to look for it, especially on days when Bonaparte's and ring-billed gulls are numerous. Their flocks scatter when a jaeger gives chase and singles out one of the gulls in an attempt to make it regurgitate its recent meal.

In the woods along the southeast corner of the lake, where the Bird Observatory maintains a banding station, small sparrows, warblers and flycatchers begin to thin out in September. But on the fields there are more passerines than ever. When a merlin hurries low over the stubble, the sky ahead fills with

Lapland longspurs and snow buntings.

By October, shorebirds have become scarce, but waterfowl remain numerous. Ducks are massed along the south shore of the lake, where no hunting is allowed and where the government has set up a feeding station as a lure to keep the ducks off the farmers' fields. The pastures along the north and west shore represent the best staging habitat for geese, but the flocks leave as soon as the shooting season opens. Several hundred Canadas, white-fronts and snow geese linger along the south half of the lake, staying as long as weather conditions allow. Throughout fall, migrating geese pass high over the lake, perhaps touching down briefly, on their way to traditional staging areas farther south.

The last major event on the birdwatcher's calendar is the arrival of tundra swans, that find abundant food and sanctuary on the lake. The babble of high-pitched voices fills the air as thousands of the stately white birds dredge for pondweeds, accompanied by ducks that pick up the scraps. The swans stay until November when winter sets in. The lake's waters cool quickly, and soon there is ice as far as the eye can see. It spells a sudden end to seven months of teeming birdlife.

The last waterfowl huddle together in water holes. Bald eagles wait nearby on the ice and make frequent passes over the massing ducks in attempts to capture the cripples. Snowy owls also kill the odd duck. If their major prey, the meadow vole, is abundant, the owls will stay all winter, but if food is scarce they fly on south.

By late November, the snow-covered lake resembles an arctic scene. Deep drifts pile up against the willows on shore. An occasional coyote ventures out of the woods, where tracks reveal the presence of deer, rabbits, weasels and the odd mink. But birds are few and far between until March, when the first waterfowl return.

Chapter 4

GREBES

When the American Ornithologists' Union organized all birds into an universally accepted sequence of orders, families and species, the scientists followed an evolutionary plan and headed their list with the loons and grebes, that were considered the lowest form of bird life, closest allied to their reptilian ancestors. In 1988, researchers proposed a new classification, based on DNA analysis, that placed the loons and grebes near the top of the list with the more advanced songbirds.

Be that as it may, it makes sense to begin the bird chapters in this book with the divers, since they are wholly of the lake. Awkward on foot and never touching land unless forced to, loons and grebes are well-adapted to spend their entire life in the water. Their wings are short, the tail is rudimentary, the body compact and encased in a layer of fat. The legs are a marvel of avian engineering. The tibia or muscled portion is out of the way inside the feathered outline of the streamlined body. The tarsis or shank is blade-like, laterally flattened for minimum resistance, while the toes are widened with long, flexible lobes for maximum pushing surface. With the heel joint in an extreme posterior position, the feet propel the bird like an outboard engine.

The plumage of divers is dense and smooth like fur. In the old days, grebes were slaughtered for their skins, in high demand for the manufacture of coats and capes, jeopardizing the survival of some species in settled regions of North America. Fortunately, after the practice was outlawed, the birds recovered, although they now face man-made dangers of a more insidious kind.

Loons, that prefer clear northern waters, are only transients at Beaverhills Lake, but several species of grebe reach high densities in the shallow and productive bays in the northeast. The western grebe is the largest and commonest, super-slim and elegant with fiery red eyes in a distinguished black-and-white face. It is not easy to get a good look at this low-slung diver, since it usually stays in the deeper water far from shore. But it calls attention to itself with its weird and endlessly repeated "kreek, kreek". Through binoculars, we can watch its springtime display and the amusing mating-dance. Side by side, body stiffly erect, the pair race on the surface like water skiers, their pattering feet kicking up a trail of foam until the birds plunge under. Several pairs may rush about at once, performing a water ballet with intricate gyrations that lead to the communal nesting areas.

Typical of all grebes, the western lays its eggs on a floating platform of debris and mud, anchored to the reeds away from shore. Like a buoy, the nest goes up and down with the changing water levels. When the incubating bird leaves temporarily, it covers the eggs with soggy weeds to conceal them from spying eyes, but on clear days the sun will keep the eggs warm enough.

Hours after the young hatch, they follow the adults into the water. They may ride on mother's back, sheltering under her wings even if she dives to pursue food. The sight of a grebe with the striped heads of chicks protruding from its feathers is surely one of the cutest summer scenes at the lake. Close-up observation reveals that the young are fed during the ride and that they are often given small feathers to swallow. Stomach contents of adult grebes, examined by ornithologists, have contained up to 70% feathers. Obviously, they play some role in the bird's digestive system, although their food value cannot be high.

Grebes are hunters of fish, aquatic insects, and crustaceans. The western and the far less common red-necked grebe pursue minnows and sticklebacks, but three smaller grebes, the eared, horned and pied-billed, subsist mainly on scuds, the shrimp-like creatures that are much more numerous than fish in the lake.

Eared grebes nest in large communities, in some years containing one or two hundred pairs, among the reed beds in the northeast bay of the lake. The horned grebe is a loner that inhabits a few bays of the lake as well as ponds and sloughs in the surroundings. In both species, the sexes look alike, with tufted crests that moult out in fall when the birds attain a dull grey winter plumage.

The third species of small grebe is the unobtrusive pied-billed, which has no crest and is dull brown with diagnostic black chin. Like all grebes, it has the ability to alter its specific gravity, either floating high like a cork, or submerging with only its head protruding. If it feels threatened, it will dive and come up secretively with bill and eyes only, enough to breathe and see, while remaining out of sight among the reeds. This habit of vanishing from sight earned it the nicknames ''hell-diver'' or ''water-witch'' in the days when the grebes were hunted for their skins. No doubt, the pied-billed's shyness helped it survive in the face of hunters as well as its natural enemies: harriers, eagles and large gulls. Baby grebes of all species are especially vulnerable to predatory fish such as pike. The absence of this ''freshwater shark'' is one of the reasons why the lake is a good place for water-birds such as the so-called lowest form.

While the horned grebe (top) is a solitary breeder, the eared grebe (bottom) is gregarious. Breeding colonies of the eared grebe may contain hundreds of nests.

Cormorants and pelicans share a small breeding island on the lake, although scarcity of fish forces the birds to commute to other water bodies for subsistence.

Chapter 5

PELICANS, CORMORANTS AND HERONS

Visiting birdwatchers from eastern Canada are thrilled to see typically western species that Albertans take for granted, such as magpies, avocets or marbled godwits. But few birds create as much excitement as the white pelican, locally common on the prairies and absent elsewhere in Canada. A first sighting of this avian giant is a spectacular event, especially if the pelicans are in flight, flapping in line-formation or soaring majestically on outstretched wings that span nearly three metres. For so large a bird, the pelican makes a surprisingly graceful impression, flying stately and in easy mastery of the air, no matter how strong the prairie winds may blow. In the vast and scenic solitude around the lake, a flock of pelicans on the wing is western wildlife at its best.

In recent times, pelican numbers at the lake have varied a lot. No doubt, they nested commonly during the high water levels at the turn of the century, unless the early settlers destroyed the colonies by robbing eggs and young for subsistence food. After protective measures came into effect, the pelican's chances improved, and they were reported nesting at the lake as late as 1950. However, from 1960 to 1974, only non-breeding pelicans were known to occur. The situation changed dramatically after the water levels rose and small fish became abundant. As many as 400 pelicans summered on the lake in 1975. The following year, they made a nesting effort. Unfortunately, the island colony was visited by several parties of birdwatchers, who may have involuntarily caused the abandonment of

the nests. Even more so than other colony-nesting birds, newly-established pelicans are very sensitive to disturbance. Their young are prone to perish if exposed to hot sun, from which they are shielded by the parents with infinite patience. In 1976, a few pelicans nested successfully, and since then the colony has had between ten and a hundred nests each year. From 1980 to 1985, the pelicans were monitored by staff from the Alberta Provincial Museum who reported a high nestling mortality of up to 80%, which was probably related to food quality. The young birds were found to be lighter in weight than the young of pelicans nesting on other lakes in the province. As indicated by the spilled crop contents of the young, the Beaverhills colony was fed primarily on aquatic invertebrates and only a few sticklebacks.

Highly gregarious, pelicans do everything together, including foraging. Swimming slowly in formation, they drive minnows, amphibians and invertebrates before them, trapping them in shallows or against stands of cattails. The birds drag their huge bills through the water, raising them at intervals to drain the contents that are collected in the deep pouch stretching from the mandibles to the throat.

When the birds return to the colony, the young thrust their bill into the pouch and throat of their parents, which regurgitate a mass of half-digested food. Blind and helpless for almost two weeks, the juvenile pelicans will not begin to fly until the end of summer, at ten weeks of age.

The statuesque pelicans have an unlikely companion on the island: the cormorant, a sleek diver in the water, but an ungainly creature on land. The two species have different nest-building habits. While the pelican lays its two eggs in a slight depression on the bare ground or on a foundation of debris scraped together, the cormorant is a compulsive builder, adding and improving each year, even stealing material from the neighbours, until the nest mound is 30-50 cm high. It consists of reed stalks, weeds and branches, cemented together with excrement. Elsewhere, cormorants nest in trees on bulky platforms that are in constant need of repair and have been known to contain over a thousand branches.

The cormorant species occurring in Alberta is the double-crested cormorant, named after the inconspicuous tuft of whitish feathers on either side of the head, displayed only during the courtship period. Unlike pelicans, which seem all but mute, the cormorant utters a variety of grunts when its passions run high. Outside the mating season, it is a quiet bird that works alone. Like pelicans, the cormorant has a throat pouch, but it is small, offering little resistance when the sinuous diver pursues fish, seizing it in the hooked bill before swallowing.

Of late years, as fish stocks declined, the lake's pelicans and cormorants have been forced to commute to better foraging locations such as Hastings Lake that contains some yellow perch. The cormorants usually travel in small parties of less than ten, flapping their narrow wings, but on warm days they gain altitude by soaring, like pelicans, before setting course for their fishing hole.

The lake's changing food resources have had an impact on another conspicuous consumer of fish, the great blue heron. A few herons can always be seen around the lake from spring to fall, but they have seldom nested there. In 1975, a small colony established itself on Dekker Island. From the opposite shore on the mainland, the tall birds could be spotted on their nests in the bare poplars, that had died

as a result of flooding. The following year, the colony increased to twenty nests, but it declined quickly again in response to the dwindling fish supply.

During the lake's productive years, another member of the heron family thrived, the black-crowned night-heron. In 1976, a total of 549 nests were counted in the drowned willows of Dekker Island, with smaller colonies elsewhere on the lake. Interestingly, this bird was reported nesting for the first time in 1959, after it had reached central Alberta in a range expansion that already included all continents except Australia, where a closely related night-heron occurs. World-wide, the heron family contains about sixty species, but only the night-herons feature a sharp contrast between adult and immature plumages. Very different from their sophisticated black-and-grey parents, first-year night-herons are a drab brown with spots and streaks, resembling another local relative, the American bittern.

For effective concealment, the bittern is not only helped by its brownish camouflage colours, but also by its compulsive habit of standing upright, long neck stretched and bill pointing skyward. Its yellow eyes are placed so that they can peer ahead at the object of suspicion. When a person walks by an alert bittern, the bird turns its head as imperceptibly as the hands of a clock, so that its frontal silhouette keeps facing the person. The streaked pattern of throat and neck blends with the vertical line of reeds, but even if the bird stands quite in the open, it usually escapes detection until it flies up at close range. When in flight, bitterns resemble immature night-herons, but the two can be distinguished by the way they hold their bill. The night-heron points straight forward, whereas the bittern's bill is cocked upward at a slight angle.

In the nesting season, the bittern is seldom on the wing, but its presence in the marsh is broadcast by its unique "song" that sounds like the hiss and suck of an old-fashioned water pump. The call is repeated frequently in series of three or four "strokes". In early spring, before the cattails have grown up, it is possible to spot the ''slough pumper'' standing on a low mound of dead plants or a muskrat house. Peeking over the surrounding marsh, the bird watches its neighbours that stand on similar platforms and emit their booming calls in convulsive hiccoughs, with the bill jerking upward at each ''pump-a-clunk''.

Hidden in the rank growth of the summer reeds, the foraging habits of bitterns and night-herons are seldom observed. They steal about on the quaking ooze on their large feet, the dagger-like bill ready to strike at a wide variety of creatures, including frogs, insects, mice, young birds and fish.

The only heron that does its hunting in the open where it can be watched is the great blue. It may stalk its prey stealthily, but it is usually passive and immobile like a fence post, albeit alert to the smallest creatures. Suddenly, its neck uncoils, and after a tense moment, its bill spears the water at incredible speed,

grasping a minnow or tadpole that has come within range. Instead of skewering the prey, as one might expect, the heron uses its long mandibles like a needle-nosed pliers, firmly holding the struggling victim. Adroitly, the bird flips the prey up and catches it head-first in the gullet before swallowing. Larger fish or mammals are squeezed thoroughly, flipped and recaptured several times before ingesting them whole. In capturing aquatic prey, herons seldom miss and are apparently capable of compensating for the deflection caused by the water's surface.

The great blue is an ancient and adaptable species that has conquered the northern hemisphere despite the briefness of the season. Before ice locks up their food supplies, herons migrate far and wide. Many will never return, succumbing to the dangers and hardships of an environment that gets more and more polluted and crowded by a far less ancient, but even more adaptable species, our own.

If it were not for its distinctive call emanating from the marsh in spring, the secretive bittern would escape notice, even when it stands quite in the open.

During migration, flocks of snow geese mix with white-fronted geese (top), and among them may be the odd blue goose (bottom), a colour phase of the snow.

Chapter 6

THE MIGRATIONS OF GEESE

Is there a more impressive sight for a birdwatcher than ten thousand geese rising massively from the shore of a lake, with thunderous wingbeats and explosive honking? As the birds pass overhead and organize in vibrating skeins, the watcher stands in awe, touched by the vitality of a spectacle that he hopes to see each spring and fall.

The gathering of migrating geese takes place at the lake twice yearly and is as predictable as the passing of the seasons, but it is easy to miss peak days; the largest flocks can stay hidden in out-of-the-way spots. Geese are particular and traditional in their choice of staging areas. At the lake, spring concentrations of geese used to occur on the fields along the west shore. But of late years, the greatest numbers collect on the remote points and bays in the northeast, which is rarely visited by people. In morning and evening, when the birds head for feeding grounds inland, a procession of flocks can be seen through binoculars from across the lake. At any time during the day when a disturbance occurs, such as the passage of a low aircraft or an eagle, the geese rise like a dark cloud over the eastern shore.

Close figures on the number of migrating geese that use the lake as a staging point during spring are not available. Rough estimates of resting flocks, flushed by government biologists from the air, are about forty thousand, but the total movement of geese over the area is difficult to guess, since the birds come and go in waves over a six to ten week period.

The earliest spring sighting of Canada geese recorded since 1965 is the first of March, about two weeks ahead of the usual date. These early-returning geese are local breeders and their arrival generally coincides with mild weather. Tied to the advance of the season, the spring movement of geese across the interior of North America is unhurried with plenty of stops to feed. It is not a helter-skelter rush as it is for other species that can be as regular as clockwork after a long-distance flight from South America. Once the geese have arrived at the lake, they are hardy enough to face a return of winter, stoically waiting out snowstorms until milder conditions return.

Canada geese range in size from the giant *Branta canadensis maxima* of the interior plains, which can weigh up to eight kilograms, six times heavier than the tiny cackling Canada (*B. c. minima*) of the Pacific northwest. Eleven separate geographic races have been recognized, although intergrading obscures the minor differences between some of them. The lake's local Canadas are a mixed western race of 5-7 kg in weight, visibly larger than the short-necked northern geese that come later in April. The first wave of migrants may spend a few days on floodwater, gleaning spilled grain from the stubble and savouring the first stink-weeds that grow in the thawing soil. They leave as soon as weather conditions allow. How-ever, many more Canadas will stay well into April and May, mixing with and eventually outnumbered by white-fronted geese.

It is always a thrill to hear the high-pitched

gabble of white-fronts. Albertans are fortunate to be on the migration route of these pretty geese, that nest circumpolar and occur in northern Europe, but do not pass over eastern Canada. Western hunters call them ''speckle bellies'', a descriptive name that should apply only to the adults, since first-year birds lack the black bars on the underside that are typical of adults. Juvenile white-fronts are always in the minority, even in fall, especially in years when spring comes late to the arctic, preventing tundra geese from breeding successfully.

Late winter conditions at nesting time also make or break the reproduction of snow geese. After a good breeding season, pairs of adults accompany three or four dusky young on the long trip to California winter quarters. Many young will not survive to go back north with the flock.

The odd snow goose may turn up at the lake in March, mixed in with the Canadas. But pure flocks of white geese arrive well into April, calling harshly, and flying in the peculiar, quivering lines, caused by the vibrating effect of the black-and-white wings, that are the origin of the name ''wavies'' preferred by old-time hunters.

Among the snows is the odd blue goose, grey with white head and belly, a colour phase of the lesser snow goose. Blues are common in Manitoba, where they may outnumber white geese, but they are rare farther west. At the lake, we may expect one blue for every thousand or so white geese. As recently as the 1950s, blue geese were considered a separate race, until breeding colonies around Hudson's Bay were found to contain both blue and white birds. Further research revealed that variations in the arctic climate at nesting time correlated with the proportion of blue and white geese in the fall population. In the early 1960s, when the tundra region experienced less severe weather than in preceding years, the number of blues increased. In the late 1960s, a cooling trend with more snowfall coincided with a decline in the proportion of blues. White geese have an earlier nesting date and are better adapted to cold spells than

blues. When spring comes early, both phases produce broods, but the young of white geese stand out on the snow-free ground and are more vulnerable to predators than blue goslings. Thus, the existence of two colour phases is believed to be an adaptation to the fickleness of arctic springs, an adaptation that makes the snow goose a very successful species. Of late years, it has responded vigorously to increased protection on wintering grounds in the United States, where government agencies have set up a chain of waterfowl refuges. There seem to be more snow geese now than several decades ago, but this is not evident at Beaverhills Lake, where they have declined markedly.

Prior to the 1960s, spring estimates could be as high as 30,000, and the flocks were said to cover fields like winter snow. Now it is rare to see several thousand together, and most flocks contain less than a hundred birds. The reasons for this decline are a matter of opinion and speculation. It is possible that the former multitudes originated from a very large Russian colony on Wrangell Island off Alaska, where disastrous declines, from several hundred thousand to only tens of thousands, occurred during the 1950s, ostensibly due to severe spring weather conditions. Another theory is that snow goose migrations have shifted from central Alberta to Saskatchewan, where large increases were recorded.

In fall, snow geese are even less numerous at the lake than during spring, although migrating flocks can be seen to fly over in late September, but they seldom come down. It seems that the lake has lost its attraction to staging geese. A possible explanation is that suitable open habitat has shrunk during the past thirty years, when formerly open pastures became overgrown with poplar and willow. Increased disturbance by people and the opening of hunting along the north half of the lake may have forced the geese to seek refuge elsewhere. As discussed in the chapter on hunting, all three species of geese have declined at the lake during fall. Wary and sensitive to disturbance, geese demand protection on their resting areas, making responsible management a necessity.

Curiously, there is one species of arctic goose which is not shy at all: the charming little Ross's goose. White with black wingtips, it looks like a snow goose, but it is smaller, almost duck-size. However, size is a poor criterion for field identification, and to tell a Ross's from a snow, the birdwatcher is required to approach closely and carefully, even if it means a belly-crawl, to get a close look at the diagnostic features, such as the stubby, unmarked bill.

Ross's geese may show up in the first half of September, well before the snows arrive, and during spring they linger well into May after other geese have left. Neither numerous nor regular, this delightful northern visitor may add a bonus to a day of geese-watching at the lake.

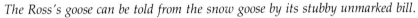

The Ross's goose can be told from the snow goose by its stubby unmarked bill.

Arriving as early as possible during a mild spell in the first half of March, local Canada geese wait out a return of winter on some open water. At the approach of people, a brooding goose makes itself inconspicuous by flattening out on the nest, hiding its white cheeks.

Chapter 7

THE LOCALLY BREEDING GEESE

The familiar grey-brown Canada goose, with its black stocking neck and the white chin strap is high on a list of bird species the visitor looks for at the lake. From March to November, it can be seen and heard anywhere along the shores. But that was not always the case.

During the 1930s, breeding populations of Canada geese had practically vanished from central Alberta. Widespread drought had reduced waterfowl habitat. As lake levels dropped, breeding islands became accessible to country folk, who augmented their diet with all the waterfowl eggs they could find, while the birds themselves were shot and netted whenever the opportunity presented itself.

Protective legislation and practical conservation efforts by government and private agencies have returned the Canada goose to central Alberta, probably in greater numbers than before. The Alberta Fish and Wildlife Division pioneered a successful reintroduction technique that involved the release of goose chicks trapped in the wild in southern Alberta. Young geese are known to return the following spring to the place where they learned to fly. Released at Joseph and Oliver Lakes, west of Tofield, the young geese migrated south from Alberta in fall, but few came back the following spring. Apparently, hunting losses were very high for these inexperienced youngsters. To provide them with much-needed guidance on the hazardous journey, Fish and Wildlife staff trapped wild adult geese in southern Alberta and released them together with their goslings at Joseph and Oliver Lakes. In fall, the young birds flew

south with the adults, and, although the adults did not return to central Alberta the following spring the youngsters did, and their survival rate had much improved. These geese became the nucleus for a locally breeding population that continues to expand and only seems to be restricted by suitable habitat and nesting sites.

The lake's geese are still under active management by the provincial Fish and Wildlife Division and Ducks Unlimited Canada, who cooperate to provide suitable nesting sites. Canada geese are very adaptable and can make a living on nearly all water, varying from farm dugouts and city ponds to wilderness lakes and rivers. But the species has one major requirement that must be met: a nesting site relatively secure from mammalian predators. Unlike ducks, a goose will not lay its eggs on the mainland, no matter how well-concealed the nest may be in grass and reeds. A goose must have a site surrounded by water or open space that allows a wide view of the neighbourhood, both for the brooding female and the gander that stands on guard at a distance. As to shape, size, building material, or height of a suitable nest, a goose is not particular at all. It will adopt small islands, muskrat houses, beaver lodges, haystacks, barns, osprey nests or cliff ledges, as long as the twin requirements of safety and unobstructed view are met. It is quick to take advantage of man-made structures, including wash tubs, automobile tires or metal drums filled with straw or wood chips. Round hay-bales, placed on the ice over shallow water before break-up, are cheap to organize but have a restricted life-span. More permanent nest sites were supplied by the Fish and Wildlife Division as recently as 1987, when 14 rock and earth islands were built along the north shore of the lake. In 1974, Ducks Unlimited constructed 23 islands in Lister Lake. Together with various platforms and haybales, the total number of artificial goose nesting sites on the lake may be close to one hundred.

The actual number of geese that use the man-made sites is not monitored on a yearly basis, nor is the total goose population of the lake. It is quite possible that the majority of Canada geese continue to use natural nesting sites such as muskrat houses. To be suitable for geese, the mound does not have to be especially high. A low house will do, as long as it allows a clear view of the surroundings. Interestingly, muskrats help to keep marshes open by cutting rushes and cattails for food and building material. The geese themselves also eat young shoots of emergent vegetation, but apparently not on so large a scale as greylag geese in Holland where they play a vital role in preventing reeds from choking shallow lakes, thus perpetuating the open habitat that geese and other birds require.

Water level fluctuations of the lake have a decisive impact on the availability of goose nesting sites. Low water, resulting in the drainage of breeding marshes, can be just as detrimental as high water that floods muskrat houses and low islands. Flooding also inundates open shoreline pasture which geese need for grazing and loafing. A lake brimful of water, that laps at the base of poplars and willows on the shore, is not attractive to geese at all.

In 1974, when Beaverhills Lake rose about one metre due to exceptionally high snow melt, many local geese were left homeless although the odd new nest site was created. A haystack in a meadow, formerly well away from the west shore, had become an island surrounded by floodwater. It attracted the attention of two pairs of geese, and for days the ganders were fighting each other in vicious wing-flapping encounters. The bend of a goose's wing can strike blows like a closed fist, while the bill can deliver a mean pinch. Honking and hissing, the victorious pair pursued their rivals in the air and forced them down for more punishment.

Competition is less fierce if there is no shortage of nest sites, and geese may nest together closely if the area also contains rich feeding grounds. However, the best nesting locations go to the first-comers, and for that reason the local geese arrive as early as the weather allows, long before the lake is free of ice. When spring break-up comes, some established geese get a nasty surprise if the waters rise and inundate their home, or if wind-blown ice-floes ride up onto their island. One day in May, a huge iceberg drifted down-wind onto the north shore, precisely hitting a small natural island where a goose was sitting on eggs. The bird got up in time and joined the gander to watch the ice bury the island. If a goose's eggs are destroyed or deserted, the pair seldom renest.

Canada geese are said to mate for life and they are potentially long-lived. Captive birds have reached ages of thirty and forty years, but wild geese seldom live beyond twenty. Long before that, the vast majority fall victim to the hunter's gun.

If a goose loses its mate, it will court again, approaching the object of its desire with head held close to the ground, neck undulating and weaving, snake-like. The affair is accompanied by much hissing and rustling of feathers. Mating takes place on the water, with the female in normal swimming position or neck stretched along the surface. The geese throw water over their backs, and in a post-mating ritual face each other, rotating breast to breast, heads held high, while the male makes a snoring sound. Presently, the pair retires to the shore to preen in apparent contentment.

The female constructs the nest, furnishing it with reed stalks, twigs and grasses. She also broods alone, taking occasional breaks to feed, bathe or preen. At her return, she drips water on the eggs, which may help to ensure successful hatching. All the while, the gander remains on guard within view of the nest. When he honks a warning, for instance at the approach of people or dogs, his mate will press herself on the nest, neck stretched flat, almost invisible except to the close observer.

Incubation takes about four weeks. The chicks, usually four to six, are precocial, ready to run, swim, dive and find food, soon after hatching. They will become imprinted with the first object they see in their early hours of life, establishing for themselves who and what they are. After half a day on the nest, mother goose leaves it for good, with her bright and perky youngsters in tow and the watchful father bringing up the rear. The larger the family, the higher the gander's status among his clan. His superior dominance allows his mate and her young to feed at the best sites with a minimum of interference from other geese.

The goslings consume pondweeds and a few aquatic insects and crustaceans. Later, the family concentrates its feeding on an open section of shoreline, where the geese clip the grass as short as on a golf course, spreading fertilizer of their own brand. If related goose families meet without hostility, the young mingle and often end up together, following the dominant pair. Gang broods of fifty or sixty goslings, accompanied by a dozen adults are a common sight on the lake during June.

The youngsters grow fast and in eight weeks or so there is little visible difference between parents and juveniles, although plumage colour is duller in the young.

During the height of summer, like all waterfowl, adult Canada geese go through their annual moult and lose all flight feathers at once. For about one month, the birds are unable to fly and hide warily in the densest reed beds. During this flightless period, the geese can be herded into traps and nets, which allows researchers to band the birds before releasing them again.

Most non-breeding yearlings and two-year-olds fly north in June to moult on some wilderness lake. After the geese regain their flying powers, they rejoin their kind on Beaverhills Lake, forming large flocks in readiness for the journey south to the wintering range.

Chapter 8

DUCKS

Central Alberta is part of the pothole belt that stretches across the prairie provinces and produces about half of North America's ducks. The sight of ducks hurrying across the western sky is taken for granted from April to October. But after five months of winter, it is a thrill to see the first birds of the season over the still barren landscape. There is a vitality and spirit about flying ducks that is especially appealing if one has not seen them for so long.

The earliest arrival date for Beaverhills Lake varies yearly between March 18 and the first week of April, depending on weather conditions and the availability of open water. Mallards and pintails vie for first place, but other species are only a few days behind: green-winged teal, wigeon, redheads, lesser scaup, gadwall and shoveler. A little tardier in arriving are canvasback, ruddy duck, blue-winged teal and the odd cinnamon teal. Diving ducks of species that do not breed on the shallow lake come and go, usually staying farther from shore: common goldeneyes, buffleheads, mergansers and scoters. The great variety of ducks makes for interesting and challenging watching. The spectrum of colours, often iridescent in the spring sun, is a delight to the eye.

Perhaps each of us has a personal favourite among the many attractive species, that include such exotic dazzlers as wood duck, harlequin and hooded merganser, all quite rare in Alberta. But one of our commonest ducks, the northern pintail, is easily the most graceful, with its slender neck and elongated tail. Its slim shape is especially evident during flight, when the narrow wings propel the bird like an arrow. High altitude travel is in the pintail's blood. Twice yearly, it migrates across the continent from California or the Gulf of Mexico to central Alberta, and many go on to the arctic coast, lifting off from the lake at sundown.

The pintails that stay behind have some time to spare before they get serious about nesting. Together with their regular buddies, the mallards, early pintails visit the stubble fields to feed on left-over grain, often accompanying flocks of geese on their twice-daily foraging flights.

Unlike geese, where the sexes look alike, ducks are highly dimorphic, and for good reason. The females are camouflage-brown, but beautiful in their own right, decorated with cryptic spots and buffy feather edgings. A female's only bright colour is on the wing patch or speculum, its design diagnostic for the species; blue-purple with white borders for a mallard, greenish-purple, edged with brown, white and black, for a pintail. Ducks seem quite aware of their plumage attributes and flaunt them to their full advantage. For instance, during mating display, a drake mallard can change his head colour from flashy green to dull black by raising the feathers. Display of speculum colours is part of the ritualized posturing that rules a duck's social life. Its function is to keep pairs together and strangers at a distance, ultimately reducing wasteful stress and aggression in the flocks.

Pair formation and courtship in ducks begins very early, often before fall migration, and continues on the wintering grounds. Upon

Dabbling ducks such as the pintail that depend on shallow, food-rich prairie water for breeding suffered disastrous declines over many years, but most species have rebounded during the 1990s.

arrival in Alberta, most mallards and pintails are mated, although they will not begin to look for nest sites until April or early May. Both partners "shop around" for a suitable spot, making frequent trips inland, the female in the lead, the drake following. They walk about warily, inspecting hummocks and depressions, with one eye on the surroundings, the other on the sky. In the life of a duck, the nesting period is filled with dangers and upsetting changes to its status. The expression "as vulnerable as a sitting duck" is true in the most literal sense. Especially the pintail has many enemies that lurk in the grass or fly overhead. The most deadly is the "duck hawk", the peregrine falcon, that can stoop out of a blinding sun and seize the duck before it has time to get off the ground.

Egg production is an exhaustive physical achievement. In ten to twelve days, a mallard duck lays an equal number of eggs, each weighing up to 50 grams, totalling more than half the bird's body weight. Once the clutch is complete, the female begins a brooding period of about 28 days, interrupted only for feeding, drinking and toilet duties. Nest failures are common, especially during the early season, when cover is short and the weather can turn wintry with snow and frost. But the principal cause of egg loss appears to be predation by gulls, crows and magpies. Fortunately, ducks renest readily and the second clutch has better chances for survival although the number of eggs is smaller. Considering the energy and time a duck invests in its nest and the risk of failure, we should always avoid disturbance. And if a brooding bird is flushed unintentionally, as often happens during our walk along the lake shore, we should take the trouble of covering the eggs with a bit of down and dry grass, in the same way as the duck herself would have done before leaving the nest of her own discrete volition. Such precaution is necessary to prevent the discovery of the nest by sharp-eyed crows or gulls, that would be quick to eat the precious eggs.

As soon as the female duck dedicates herself completely to the task of brooding, her beau suddenly finds himself alone after many months of inseparable companionship including daily copulation. Some drakes seem to have trouble adjusting to their forced celibacy, since they will pursue any unescorted female that shows herself, including those temporarily away from the nest. To escape an aggressive drake, the duck is forced to take to the air, exposing herself to other drakes that fly up in pursuit. A dozen or more may eventually take part in the chase, jockeying for position behind the harassed female that quacks plaintively and tries to throw her unwanted suitors off by twisting and doubling back. Mallards can be relentless until the duck is finally overtaken and overpowered, although most chases end less violently and simply dissolve after some time.

These so-called "rape flights" are common during May, among mallards as well as pintails, gadwalls, lesser scaup and blue-winged teal. Whether or not the drakes involved are actually former "husbands" that have lost their duck to brooding duties, or unmated and frustrated bachelors, is a question that can only be resolved by intensive field research with marked birds.

After the drakes lose their sex drive, they draw together in separate flocks, until they moult their nuptial colours and resemble the females. The males of some species, particularly canvasback, fly in from far and wide to congregate on the lake and feed on the rich sago pondweeds that grow during the warm summer days.

Sometime in June and July, well-hidden in the growing grasses, mother ducks become aware of the new life breaking out of the eggs. They keep their offspring in the nest for half a day or so, before guiding them to the nearest water. Head held high, the duck is the picture of vigilance. The risk of attack by predators increases with time spent travelling overland. A duck that nests far from the lake, perhaps due to a lack of suitable nesting cover on overgrazed shoreline pasture, jeopardizes the survival of her brood on the long trek to water. Once the duck family has reached the

The Canadian Wildlife Service has captured and banded thousands of ducks at the lake. After the summer moult, species such as teal can still be recognized by their "colour-coded" wing patch.

Odd and colourful in looks and behaviour, the ruddy duck is one of many western waterfowl species that do not occur in eastern Canada and that visiting birdwatchers come to see at the lake.

lake, the young may be safe from land-based predators, but large gulls remain a threat. They swoop down over the reeds before the duckling has a chance to dive and before the mother can rush in to defend her youngster.

If the ducklings survive the first critical weeks, they grow to be as large as mother in about two months. By that time she enters the annual moult. Ducks spend this flightless and vulnerable period in the densest reed beds. About four weeks later, the family is ready to try its wings and explore the countryside for a change in diet: the ripening grains on the farmer's fields. Once the ducks have zeroed in on a favoured location, their number and depredations grow until it constitutes a serious loss to the landowner.

In the past, government agencies used to reimburse farmers around the lake for a portion of waterfowl crop damage, paid for with money tacked on to hunting licences. But since 1971, the Fish and Wildlife Division and the Canadian Wildlife Service have concentrated on preventing grain losses by luring the ducks to bait stations on the south and east shores. A local farmer is contracted to dump a truck load of barley each day until 80% of field crops has been harvested. The station on the east shore was flooded out in 1974, but the fenced and posted feedlot on the south shore has been in operation each summer. It proved to be very effective in concentrating the ducks and keeping them off the fields.

When the massed mallards and pintails rise at the approach of people, their number seems incredible, and we wonder whether there really are less now than formerly. Comparative figures for the lake's duck population are not available, but estimates for southern Alberta show a steep downward trend since the 1950s, when almost two million each of mallards and pintails were believed to nest. By 1986, they had dwindled to an estimated 700,000 and 300,000 respectively, with further declines to record low levels in 1988.

A major cause of this drastic decline has been the accelerating loss of nesting habitat due to agricultural practices and drainage.

Potholes and sloughs, that once dotted southern Alberta in a mosaic of water and land, are especially important to breeding ducks. The shallow basins, often temporarily filled after snow melt, warm up quickly in spring and soon teem with aquatic crustaceans and insect larvae that supply ducks with a high-protein diet that is vital for egg production.

Formerly, most sloughs were surrounded by wide margins of undisturbed natural vegetation in which a duck could hide its nest with a fair chance of escaping detection by sharp-nosed predators such as coyotes and skunks. These critical vegetation zones have been narrowing over the years as farmers bulldozed, burned and plowed up to the water's edge, assisted and encouraged by government grants and financial incentives for acreage brought into production. Once the willows and reeds were gone, the pothole's evaporation rate increased, and in dry years the farmer got a chance to plough and cultivate the entire depression.

The accumulating loss of wetland habitat in southern and central Alberta may have forced ducks to concentrate on Beaverhills Lake where declines seem less noticeable. The massive build-up that occurs each August and September at the feeding site has been estimated at twenty to thirty thousand birds. They were fed between two and ten thousand bushels of grain each year from 1971 to 1987.

Before farmers and government agencies began feeding ducks on such a large scale, on the fields as well as at the bait station, the armies of dabblers had to rely on natural foods, mainly aquatic plants and their seeds. Collectively, the ducks must have played a role in the distribution of these plants, a role that was terminated or reduced after settlement of the surrounding land. German researchers have reported that ponds with numerous ducks had superior water quality because the birds speeded up the recycling of nutrients that benefitted fish and fish-eating creatures. Such a ''food processing'' function of waterfowl should indeed be reduced if the birds have unlimited access to grain. This intriguing point may be of minor importance at places such as Beaverhills Lake that is too shallow for fish anyway. Another concern often raised in connection with waterfowl management practices that concentrate birds in dense pockets, is the increased danger posed by contagious diseases such as avian cholera. Recent epidemics on refuges in the United States, where waterfowl are massed from fall to spring, have given cause for worry. Infected birds that migrate north may die along the way, spreading the disease to other staging points. Thusfar, no cholera outbreaks have occurred at the lake. But in some years, large-scale duck mortality is attributed to acute botulism poisoning. The bacterium that releases the fatal toxins is endemic in the lake and multiplies explosively if the water temperature rises over the shallows. Ducks and other birds that eat infected snails become paralysed and lose the ability to fly and to stand up, or even to raise their head. Typically, the nostrils become clogged with leeches. After the birds die, the toxin concentrates in the carcasses, creating a reservoir for future infections.

Formerly prevalent only in the American west, the botulism bacillus of the type C variety has recently been identified in Europe, where industrial effluents have raised water temperatures. In densely-populated Holland, soon after an outbreak is reported, armies of volunteers assist biologists with the collection of crippled birds, which are rehabilitated to a clean environment. Dead birds are destroyed by incineration.

During epidemics at the lake, waterfowl agencies collect botulism victims along some sections of shore, but the lake is too large for complete coverage. No doubt, in the past, large carnivores such as bears provided an useful service in cleaning up carrion, a task that now is left mainly to coyotes and scavenging birds. The intestinal tract of most scavengers should be impervious to the botulism toxins, but some individuals appear to be vulnerable. If the outbreak persists,

crippled crows and gulls, as well as the odd harrier and short-eared owl, will be grounded among the ducks and shorebirds they fed upon. Even falcons are at risk. At the lake, peregrines and merlins easily capture prey that is affected by botulism. During outbreaks in Utah, peregrines were found crippled themselves. Fortunately, recovery is possible if the birds are supplied with clean food for some time.

It is a great relief when cooling temperatures and heavy rain put a stop to the insidious plague. For weeks after, the remains of ducks continue to attract scavengers, until well into fall when bald eagles arrive from the north. Always quick to accept a free lunch, the regal birds utilize even the most putrid carcasses, besides the more recent victims of the shooting season that started in the middle of September.

As soon as the lake echoes to the thunder of shotguns, waterfowl vacate the most popular hunting spots and concentrate on the south half of the lake where no shooting is allowed, and where a Canadian Wildlife Service crew often bands ducks. They are caught in wire-netting traps that are about five metres across and may hold as many as a hundred ducks. Baited in with barley, the birds are unable to find their way out, until they are released by the biologist, who checks the traps once a day. In their panic to escape, a few ducks may injure themselves against the wire. Others are less stricken with fear and re-enter the cage several times after release.

Between 1975 and 1986, the C.W.S. crew captured 5338 ducks, mostly mallards and pintails. Thus far, 368 (6.9%) were reported recovered after release. Across North America, since 1908, nearly ten million ducks have been banded, of which 1.2 million were recovered at distant locations, representing a return rate of 13%. The data have given researchers a useful picture of duck migrations across the continent.

An interesting bit of incidental information that has come out of the banding effort at the lake, is the unexpected occurrence of black ducks, an eastern species that replaces the mallard in Atlantic regions and in Ontario. Male black ducks resemble female mallards and are tricky to identify in the field. Spotting one of these odd ducks would be a highlight for the birdwatcher who comes to the lake to take advantage of the crisp fall days and to enjoy the variety of waterfowl, before they fly south and leave Alberta's big sky empty for five long months.

A canvasback draws her ducklings close, guarding them against predators. Especially the larger gulls present a constant danger and may swoop down over the reeds to carry off a baby duck before mother can rush in to defend her brood.

In the field, immature bald eagles (above) are tricky to tell apart from golden eagles. When soaring and seen from below, the wing shape of the two species is markedly different. The trailing end of the golden's wing curves inward near the body, whereas the bald's wing is quite straight and rectangular. The bird at right is a golden eagle.

Chapter 9

EAGLES AND OPEN-COUNTRY HAWKS

Of the 18 species of diurnal raptors recorded in Alberta, only three are known to breed commonly around the lake. Another three nest on an irregular basis. However, during migration all eighteen have been seen and can be expected again, making correct identification often difficult or impossible. Because of the close similarity between related species, our eagles, hawks and falcons are doubtless the most frequently mis-identified group of birds in the field. The problems are many. Shy and often on the move, birds of prey seldom allow more than a passing glance. The flight silhouette of most species is similar with short neck, prominent tail and large wings, that differ only in subtle proportion. Size is relative and of little help in the field, especially in view of the overlap between species and the size-dimorphism between the sexes. And thirdly, the colour scheme of many raptors is a sub-dued brown and grey with separate plumages for immatures, and confusing variation between individuals of the same species.

To illustrate the pitfalls of bird identification, Roger Tory Peterson, America's foremost birdwatcher, related an anecdote in which three groups of people on an Audubon Christmas Bird Count briefly stopped to identify a road-side hawk. When they reassembled later, each party reported a different conclusion, to everyone's con-sternation! Unfortunately, birds of prey are often (mis)identified on the basis of impressions instead of facts, or on a single perceived characteristic, when it was actually not possible to see any of the truly diagnostic features. For instance, distant immature bald eagles are often reported as golden eagles because the bird in question *looked* quite dark. Immature peregrines have been mistaken for gyrfalcons because they *seemed* too large for a peregrine. And immature red-tailed hawks, with their whitish undersides and dark marks on the belly and in the wings, are frequently mistaken for American rough-legs. In each case, the observers jumped to conclusions, prompted by a compulsive desire to label living things. Only the best birders are not afraid to admit that their observation did not allow for positive identification, since all they saw was a dark eagle, a large falcon, or a distant light-phase buteo.

Fortunately, a few of our raptorial birds are unmistakable at a glance, such as the magnificent adult bald eagle, with its white head and tail, contrasting sharply with dark body and wings. It is an impressive sight to see it fly along the lake shore, creating panic among the ducks and geese, that rise ahead to hurry out of the eagle's flight path. Coots and diving ducks, slow on the take-off, may stay behind and rely instead on their ability to submerge if the eagle makes a pass at them. An eagle that really means business will fly in tight circles over the diving prey, or hover low over the water if the wind is strong enough. As soon as the coot or duck comes up for air, the eagle swoops with its taloned feet ready to grasp the prey. In clear water, it can see the submerged target and anticipate its return to the surface. In a flash, the eagle seizes the prey and carries it to a plucking spot.

The two most numerous hawks around the lake are the red-tailed (top), which nests in trees, and the harrier, or marsh hawk, which raises its young on the ground in reeds or buck brush (bottom).

In late fall, when the lake is frozen over and diseased or crippled waterfowl are concentrated in water holes, kept open by the birds' activity, bald eagles congregate nearby. As many as four or five, or even a dozen, can be seen standing on the ice. To watch them hunt, we need to have patience. If it is too cold to stand or sit in the open, we can observe from a vehicle and look through a telescope. Eventually, one of the eagles will make a pass over the ducks that mass together tightly. They dive and splash if the eagle swoops low, but more often than not, the big bird will hesitate and resume its vigil. Ducks that clamber onto the ice for a nap are certain to attract attention. The eagle approaches hurriedly and snatches the prey in one foot just as it reaches the water.

Of 118 attempts at capturing ducks, the author has witnessed 14 kills, a success rate of 11%, quite similar to that of the peregrine falcon. Apparently, despite its reputation of being lazy and content to feed on carrion and fish, the bald eagle is a capable duck hunter. During migration and on the wintering grounds, it has a semi-social lifestyle and a food-sharing system based on the principle that might is right. The dominant and hungry eagles rob the weaker ones, that wait their turn and fight over the remains. It is not unusual to see a freshly-caught duck change "hands" three or four times.

As long as there are ducks to hunt, bald eagles remain in central Alberta until well into December. Evidently, when south-bound, they are not in the same hurry to reach their destination as they are on their way to the northern breeding grounds. Over twenty years, from 1964 to 1983, the author counted 626 sightings in fall as opposed to 187 in spring, with the number of observation days consistently greater in spring than in fall.

The first south-bound eagles may show up in mid September, but the peak days are as late as the second week in November, with adults arriving and departing two to three weeks after the immatures. The proportion of adults in fall is about one for every immature,

but in spring there are two or three immatures for every adult sighted. One would expect the opposite, less immatures in spring, considering the high first-year mortality suffered by all raptors. The odd spring ratio can be explained as a result of differential migration routes for the two groups. On their way north, eagles of breeding age avoid central Alberta and choose a flight path along the mountains, where the topography favours the formation of thermal currents, which the birds seek for ease of travelling. In the western foothills, adult bald eagles are numerous in March, three or four weeks before immatures and a few adults are seen passing over the lake.

Also the migration of golden eagles is more pronounced along the mountains than it is in central Alberta where they are quite uncommon. From 1964 to 1983, the author recorded only 18 in spring and 14 in fall, for a total of 32. Compared to the 813 bald eagles seen during the same time period, the sighting frequency of goldens to balds is one to twenty-five.

Golden eagles are difficult to recognize with certainty from a distance if the "golden" nape is not evident. Adults are dark with varying amounts of bleached-brown on the shoulders. Immatures show white flashes in the under-wing at the base of the primaries, and they have a whitish tail, broadly edged in black. The amount of white decreases as maturity approaches, a process of three or four years.

The two species of eagle differ in bill size and leg feathering, which is not obvious in the field, but wing silhouette can be an excellent help in identification. The golden's wing tip is wide and rounded, while the base of the wing narrows near the body. By contrast, the bald's wing tip is angular, with the first three or four primaries longer than the others, and the wing does not narrow at its base. The difference can only be appreciated if the bird is seen in full silhouette with wings spread in soaring position.

The plumage of immature bald eagles is variable. Most show areas of white or cream colour in the underwings and on the body, but a few are uniformly dark or mottled. In the last stage before full maturity is attained, the tail moults into white, resembling the immature golden.

The inherent difference between the two species of eagle is reflected in their behaviour. The golden has no particular affinity for the lake, but it may on occasion fly along the shore and harass ducks and geese. It is capable of surprising bursts of speed when it tries to seize flushing mallards in mid-air. But most commonly, it ranges over the land, and its staple prey is rabbits and ground squirrels, with the occasional muskrat or even a fox. One spring day, an adult golden was seen to descend in a low area along the west shore where swarms of meadow voles were being flooded out of their burrows by a sudden melt of winter snow. The rodents were captured eagerly by dozens of screaming gulls and crows, as well as by several bald eagles, before they were joined by the golden eagle. Apparently, if given the opportunity for an easy meal, even the "king of birds" is not above stooping to the lowliest fare.

Voles are usually the mainstay of smaller birds of prey, especially the northern harrier. Formerly called marsh hawk, the harrier is the most common raptor around the lake. Already in March, before the snow is gone, the first harriers can be seen over the pastures and fields, quartering the ground in their slow manner, searching for voles and mice. The adult male is gull-coloured, in contrast to his mate which is brown with a prominent white crescent on the base of the tail. If the pair finds food in abundance, romance begins early and is announced by both partners in the sky-dance, a display flight of elegant u-shaped dives and ascensions, either high in the air or low over the reed beds. The hawks keep up a staccato commentary, and flip up-side-down like stunt flyers. This raptor's skill is also evident during the exchange of food between the two. When he arrives with a small prey in his feet, she rises from her nest in the reeds and approaches the flying male from behind and slightly below. He drops the food and she catches it in mid-air.

In years when mice and voles are scarce, the harrier's nesting success is jeopardized, since the female may have to forage for herself. Larger than the male, she kills the occasional small duck. One hungry female tackled a coot, but had to let it go again after the victim's mate rushed the hawk that was standing belly-deep in water holding its submerged prey. The male may also switch to birds and abandon his usual slow flying pace for an accelerated coursing strategy, designed to take ground-dwelling sparrows and larks by surprise. One adult male caught a pectoral sandpiper. When a red-tailed hawk approached with piracy on its mind, the harrier flew up quickly, leaving its prey in the long grass. Unerringly, it returned to the exact spot, after the red-tail had given up its search and left.

Birds of prey are keenly aware of each other and food theft is common, both intra-specific and inter-specific. The strong and fast take from the weak and slow. Even the peregrine falcon is not above robbing the harrier. But the worst kleptomaniacs that steal from the peregrine as well as from the harrier are the buteo hawks, powerful and slow, but with a keen eye for a free lunch.

The most common buteo around the lake is the red-tailed hawk. The adult is one of the few raptors that can be told at a glance. In good light, its brick-red upper tail can be seen from a long way off. However, in some races and individuals the colour is more whitish than red. Immatures and yearlings have brown tails, finely barred with black or grey, without a hint of red. How can we distinguish these non-red red-tails from the other locally breeding buteo, the Swainson's hawk? As was the case with the eagles, a good field mark, not emphasized in the handbooks, is wing-shape. The red-tail's wing is full and rounded at the tip, whereas the Swainson's primaries are pointed and angular. In addition, when soaring, the latter holds its wings at a slight upward angle. Of course, typical Swainson's hawks can be told easily by the light underwing coverts contrasting sharply with the dark flight feathers, a pattern that is

Fall migration of rough-legged hawks takes place in October and November, although a few may show up in late September. The earliest reliable report is 14 September 1974, the same year when snowy owls arrived earliest. The return migration of rough-legs lasts well into May.

reversed in other buteos.

Both the red-tail and the Swainson's are expert hunters of voles and ground squirrels, but if small mammals are scarce, they turn their attention to other fare. In low surprise attacks, launched from a fence post, the hawks take sparrows and blackbirds in the grass. During heavy rains, they feed on earth-worms and pocket gophers that emerge from flooded soil. A red-tail was seen to kill a pintail duck on the nest, and a Swainson's captured a sora rail by pouncing into a weed-infested slough. The hawk sank up to its chest into the water and rested awhile before it flew up, carrying the prey.

The identification of the above two common buteos can be compromised by the rare local occurrence of a southern species, the ferruginous hawk. In typical colour phase, it is nearly white on belly and tail, with light flashes in the upper wings near the base of the primaries. Wing flashes are also characteristic of immature red-tails and of a fourth open-country buteo, the rough-legged hawk, a transient from arctic tundras. On its return from the north, it seldom shows up before October, although many rough-legs are reported erroneously in August and early September by people who confuse them with immature red-tails. From below both hawks show dark markings on the belly and near the bend of the wing, and both share a habit of hovering, although the rough-leg is more addicted to this aerial mode of hunting because of the absence of tree-perches in its tundra homeland.

To further complicate field identification of buteo hawks, all four species have melanistic or black-phase individuals that are nearly inseparable in the field. Black buteos seem especially prominent in November, when the last of the open-country hawks pass by on the way south. To be honest with ourselves, we will just have to let most of them go as U.F. Bees - Unidentified Flying Buteos.

Adult red-tailed hawks are easy to recognize, but immatures, that lack the diagnostic brick-red tail, are often mis-identified and reported as rough-legged hawks.

Melanistic or black colour phases occur in all species of buteo hawk. The Swainson's (opposite page) can be identified while in flight by its pointed wings that are held in a shallow V as the bird soars.

Chapter 10

FALCONS AND ACCIPITERS

Before the invention of the shotgun, royalty and the common man alike sought outdoor entertainment with nature's own game-getters: falcons and hawks. Falconry is still called the "Sport of Kings", but its modern-day enthusiasts come from all walks of life and number in the thousands in North America alone. However, we do not have to shackle the falcon to capture its magic. On the contrary, to see a wild bird in aerial pursuit of prey is the greatest thrill of all, and it is available for free to the watcher at the lake.

The celebrated peregrine falcon, renowned as the fastest living creature on earth, has disappeared as a breeding bird from all of rural Alberta, but northern falcons still migrate through in fair numbers. During 15 years of field surveys, from 1969 to 1983, the author recorded 880 sightings in spring, and 51 in fall. These lopsided seasonal totals leave no doubt that spring is the best time to see a falcon at the lake. But why are they so much scarcer during fall? It is possible that most southbound peregrines choose a flight-path along the coasts of North America, whereas the spring migration takes place via a more direct route through the heart of the continent. A similar differentiation in seasonal flyways is known for shorebirds.

The earliest falcons return from their wintering quarters in the latter half of April, but the main migration period, when 80% of the author's spring sightings took place, is between May 4 and 23. Despite their preference for open habitats, peregrines seem to be overlooked by most local birdwatchers who seldom report the species. The reason is that these people typically peer through a telescope at sandpipers or ducks by the shore, while they fail to spot the falcon that soars high overhead or sits on a fence post some distance away.

Our chances of finding a peregrine are enhanced if we know its habits, permitting us to anticipate when and where it is most likely to turn up. As is the case with most migratory birds, the passage of falcons is subject to fluctuations that depend on weather conditions. Good peregrine flights, in spring as well as fall, are associated with overcast skies or rain, when most birdwatchers stay at home. By contrast, few falcons arrive during warm and sunny periods when an atmospheric high pressure system is centred over Alberta. The sinking air mass forces winds out to the edge of the system, where they turn in a clockwise direction, caused by the rotation of the earth. Winds around low pressure areas are deflected anti-clockwise. Hawks and other long-distance migrants hug the weather fronts and travel down the wind. On exceptional days, just before a depression settles in with major rainfall, six or seven peregrines may fly by in a couple of hours. In late afternoon or evening, two or three can be found resting on fence posts. Next day, soon after the weather clears, these falcons will resume their journey, except the odd one that may linger for a few days or even a week if prey such as shorebirds is abundant.

Like most of nature's hunters, the peregrine is a bird of leisure that spends much of its time at rest. In the absence of high perches

The best time to look for peregrines at the lake is between 4 and 23 May, especially when and where migrating shorebirds are numerous. The earliest spring date for peregrines is 16 April; the latest reliable fall date is 4 October.

The locally breeding population of merlins seems to have all but disappeared, but non-breeding merlins can still be seen from April to November. They like to sit on a snag or post, on the lookout for small passerines or sandpipers. If given the chance, the much bigger peregrine (top) will try to rob the merlin of its prey and it may even pursue the smaller falcon with the intent of capturing it.

around the lake, the falcon chooses a fence post or stone. During strong winds, it will sit on the ground. To avoid scaring it away, we should keep our distance and observe through binoculars or a telescope until it takes flight of its own accord. Only then will we have a chance of seeing it hunt. The wait may be long and frustrating. We may lose sight of the falcon for a moment and find it gone later. Or, after hours of patient observing, the bird may spread its wings and leave the area without any attempt at hunting. But if we are finally lucky, the joy of watching the peregrine accelerate in pursuit of prey will be all the more appreciated. Another, perhaps less boring way of waiting for a hunting peregrine, is to stay near concentrations of their prey in an area where we have seen falcon activity before. During May, when sandpipers collect by the hundreds, even thousands, on lake-side pastures, we can sit quietly nearby, enjoying the scenery and watching the birds. When an alert godwit or killdeer cries alarm, and sand-pipers rush into the sky, look beyond the fleeing flocks and search for the falcon, that has come down like a bolt from the blue. By the time we discover it, the kill may have been made already. But more likely, the hunt failed and the falcon flies up to set its wings for soar-ing. Keep it in the binoculars as long as possible, for it may make another attack not far away, giving us a ring-side view of the action.

From 1965 to 1987, during thousands of hours of patient and persistent watching, the author has seen peregrines attack potential prey species on 1478 occasions. The outcome of 412 hunts was unknown, but the remaining 1066 attacks had a success rate of 7.5%. In total, 80 prey were captured, including 50 shorebirds of 11 species, 25 ducks of 8 species, 3 small passerines, 1 gull and 1 tern.

Compared to the average shotgunner, a success rate of 7.5%, representing one hit and fourteen misses, seems ridiculously low, but it is quite in line with figures reported in studies of other predators, such as eagles, wolves and lions. The speed and agility of nature's hunters are matched by the hunted, so that the predator can only make a living by exploiting the weaknesses of its prey. It has been claimed that the peregrine is an exceptional predator, capable of capturing any bird if given a fair chance. In reality, it does not often get that chance, since prey routinely escape by plunging into cover. Moreover, like most predators, the well-fed falcon takes what comes easiest. Difficult chases are abandoned. Only if needed, will the falcon exert itself to the utmost, pursuing its victim with astonishing power and determination.

The highest hunting success rates, between 25 and 35%, have been recorded for peregrines that specialized on prey in vulnerable circumstances, such as land birds crossing lakes and rivers near falcon nesting cliffs. These breeding peregrines foraged with maximum efficiency since they had to feed not only themselves but also a mate and/or young. At the lake, migrating peregrines can afford to take life easy. Adults had a success rate of 9%, while first-fall immatures scored a hit only 5% of the time. By spring, the success rate of immatures had improved to 7%.

Adult and immature falcons use different hunting styles and strategies, which is apparent during September, when a few peregrines of both age groups come through, especially on overcast days just before or after a cold front has passed. A good vantage point to wait for these migrants is the southeast bay. Sitting on the dam and frequently aiming the glasses to the north, up the long stretch of east shore, we may discover a falcon far away, before it begins its long-distance attack on shorebirds that have congregated in the bay.

Approaching in an oblique descent, at tremendous speed, an adult peregrine got very close to a feeding flock of dowitchers that flushed in the nick of time. Overtaken at once, one of the flock dropped back into the shallow water to dodge the attack. The peregrine shot upward, stalling, reversed direction and stooped down in an instant, seizing the dowitcher just as it tried to get away.

Such effective and breath-taking action is in sharp contrast to the performance of fall immatures. One young peregrine, in its brown and streaked juvenile plumage, approached fast and low over the reed beds west of the dam. Its sudden appearance over the bay sent dozens of shorebirds shrilling into the air. One lesser yellowlegs was hotly pursued, evading several vigorous swoops, until it plunged into water. It attempted to find safety at the base of a fence post standing in the shallows. The falcon perched on top. At intervals, it made a series of about 50 passes at the yellowlegs that dodged neatly by fluttering aside, until it made a fatal mistake and jumped right into the falcon's clutches.

Young falcons do not lack speed to overtake their prey, but they often fail at the critical moment when the prey dodges. However, what they lack in expertise and strategy is made up by perseverance in long, exhaustive pursuit.

Preferring the quick and easy way, adult peregrines utilize a basic strategy of surprise, attacking the prey while it is on the ground or in shallow water. Such surprise attacks made up 74% of all hunts seen. Of 66 prey caught, 40 were seized at once, just as they flushed. Even large prey such as ducks were sometimes captured on the ground or in water and killed on the spot, after a wing-flapping struggle. These "down-to-earth" hunting methods may seem unusual to those who have read the standard accounts of peregrines knocking prey out of the sky with a blow from the taloned feet that sends the victim crashing to earth, either dead or mortally wounded. Such drastic techniques have been acquired by trained falcons that work in cooperation with people and dogs, who flush partridges and grouse below the stooping falcon. Wild birds have to create their own opportunities and seldom knock flying prey from the sky, grasping it in the feet instead and bearing it down to the ground, where it is plucked and consumed.

Of the 80 kills observed at the lake, only one was an obvious "knock-down". In a long, oblique stoop from a high soaring position, an adult falcon approached a flock of teal head-on and from below. It swooped upwards at one

of the ducks and apparently struck it with the talons, since the victim fluttered to the grass as if it had been shot. In other observations when the prey was missed, it only *looked* as if an aerial hit had occurred because the prey plunged into water or onto the ground. Such escape tactics are routinely used, causing the falcon to overshoot the mark. Often the prey takes off again immediately into a different direction, while the falcon turns around in hot pursuit.

No doubt, many reports of falcons knocking prey out of the sky are based on erroneous interpretation of such common escape behaviour. For instance, while crossing the lake, peregrines often make threatening passes at flying ducks that splash down in a split-second manoeuvre to avoid capture. It *seems* that a collision occurred between the two, and it is easy to draw the false conclusion that the duck was killed. Whether or not the falcon made a serious try at capturing the duck is impossible to tell. Perhaps, it was just playing or fooling around. However, for a predator, it is a thin line that separates play from work, and the ducks refuse to take any chance!

The most deadly peregrines often appear deceptively relaxed and indolent, soaring idly, it seems. Generally, peregrines are not considered soaring hawks like buteos or eagles. But at the lake, falcons spread their long wings at every opportunity, not only on warm and sunny days, but also during cloudy and windy weather and even in light rain, when they join the gulls high in the stormy clouds. Surprisingly, these soaring falcons are often looking for prey on the ground, and may attack in a near-vertical stoop. The use of soaring flight as a hunting method had not been reported before in the scientific literature. Neither was it known that peregrines soared during migration. At the lake the author kept many falcons under observation in early morning when they sat for hours on fence posts. Around 10 or 11 a.m., after the sun had warmed the earth and allowed the formation of thermal currents, the falcons would fly up and begin to circle, gaining tremendous altitudes where they could barely be seen in binoculars. Eventually, they set their wings and glided swiftly northward, apparently leaving the area on the way to their ancestral destination.

Another surprise was the variety in plumages among individual peregrines, a fact that is not adequately explained in the birding field guides. Dorsally, on back and upper wings, adult falcons range in colour from blue-grey to brownish-grey or blackish-brown. Their white chest distinguishes adults from fall immatures, that are heavily streaked on the entire ventral surface. However, at their return in spring after a winter of southern sun, most immatures have a pale chest and a lightly marked belly. From a distance, these bleached spring immatures resemble adults. Some young falcons, possibly belonging to the northern subspecies *tundrius*, are dorsally of a sandy colour instead of the usual dark-brown. Moreover, in contrast to the prominent malar stripe featured in the typical peregrine, some northern falcons are practically blond-headed with a narrow malar stripe, reminiscent of the prairie falcon or some gyrfalcons.

Prairie falcons, a species of arid southern regions, have been sighted at the lake in any month from late March to early November. They can only be recognized with certainty by the black wingpits or axillars. Arctic gyrfalcons have been recorded from the third week of September into November. They are northbound again before the middle of April. Extremely variable in colour from white to blackish, some individuals closely resemble peregrines or prairie falcons, while the long tail and fairly broad wings suggest a goshawk. As cautioned in the preceding chapter, the prudent watcher should be prepared to let a distant falcon go as unidentified, unless a detailed view can be obtained of all characteristics, not just an impression of size and colour.

Interestingly, prairie falcons at the lake hunt shorebirds in the same headlong manner as peregrines, stooping down low over the shore to take them by surprise. More commonly, the prairie's target is blackbirds, starlings or other open-country passerines, in addition to its summer staple, the ground squirrel, which is hunted in long-range stoops from high soaring flight.

Gyrfalcons have been seen to attack ducks. Unfortunately, these grand northern raptors do not seem to stay around long. Over the years, despite the great amount of time spent searching for falcons, the author has seen no more than a dozen gyrs in fall and only two in spring.

Much more numerous than their large cousins, the smaller kinds of falcon are easy to identify, although the large local race of merlin is sometimes mistaken for a peregrine or prairie falcon. In colour and head markings, the merlin indeed resembles a prairie, but the former's banded tail and absence of black wing-pits should identify it when seen in good light.

Merlins are specialized hunters of small birds. Of 354 hunts observed by the author, 44 resulted in a kill, including 28 sandpipers of 8 species, and 16 passerines of 5 species. Overall, these merlins had a success rate of 9%, but adult males on breeding territory achieved 38%. Adult males can be told by their blue-grey dorsal colour, as opposed to the brown-backed females and immatures.

Unfortunately, the locally breeding population of merlins has all but disappeared. Formerly, there were three or four nesting territories in woodlots off the north and west shore, but since the 1980s not a single nest has been found in that area. The cause is not clear, but may be tied to agricultural practices that directly or indirectly affected the merlin's productivity or prey base. Small birds such as horned larks have become scarce on the lake-side pastures that are not as heavily grazed as formerly, resulting in longer grass that suits neither the lark nor the little falcon. If open-country sparrows are chased by merlins, they quickly drop into the vegetation and hide effectively from the hovering and searching falcon. The overgrown condition of much of the lake shore, or the poor nutrient content of the water itself, may have caused a progressive

Captured for banding purposes, this kestrel poses for a portrait, albeit involuntarily, showing its striking head pattern.

scarcity of small shorebirds, with negative impact on the lake's attractiveness for merlins.

However, from April to November, non-breeding merlins can still be seen in the area, sitting on fence posts or flying low over the pastures and fields. Their hunting strategy is quite similar to the peregrine's; 72% of all merlin hunts were surprise attacks on prey initially on the ground or in shallow water. But unlike peregrines, which launch surprise attacks from great distances, often over one kilometre and from very high soaring flight, merlins begin their surprise attacks from less than half a kilometre, and seldom from a soaring position. Typically, a merlin starts from a fence post and flies very fast directly towards feeding sandpipers or small passerines. Success can be instantaneous, especially if the little falcon's approach was screened by tall grasses. If the prey escapes the initial attack, the merlin may give up at once or chase it out far over the lake, attempting to outclimb it and use the superior height for a downward rush at the prey. If it dodges again, the procedure is repeated, with the merlin ascending and descending in zigzags until it either abandons the attempt or succeeds in capturing its prize. If hard-pressed, passerines drop steeply back to earth and plummet into cover like a falling stone. Sandpipers often splash down into water. Adult merlins have little trouble in plucking the prey from the surface, but immatures usually fail to connect, giving the prey a chance to take off again into the opposite direction. Exceptional chases last many minutes and cover several kilometres, with the prey and the merlin climbing higher all the time. The speed and dexterity displayed in these life-and-death tests of fitness makes for animated watching, whether we identify with the hunter or the hunted.

Far less energetic and predatory than the merlin is the pretty and colourful American kestrel. It may kill an occasional bird but it usually hunts for mice and voles by hovering at some height over the ground. In the absence of suitable tree cavities for nesting, kestrels are not known to breed near the lake and occur only as transients. Travelling family groups are a common sight in August, when they feed on grasshoppers and dragonflies. They are often joined by immature merlins that also hunt dragonflies and catch the erratic insects with great skill.

Active and playful, young merlins initiate games of tag with the kestrels, as well as with magpies and crows, chasing them in and out of trees and around fences. If the merlin meets a sharp-shinned hawk, it finds its match in aggressiveness and agility. Similar in size and both expert aerial hunters of birds, the interaction of merlin and sharp-shin gives us a chance to compare the differences between the long-winged falcon and the short-winged accipiter. The falcon is an open-country hunter, capable of sustained speed, whereas the accipiter is a sprinter with rounded wings and a long tail that facilitate quick starts and turns, a vital asset in a treed environment.

In North America, the accipiter family includes three species: sharp-shinned hawk, cooper's hawk and goshawk. The latter is the biggest and a true woodland hunter that visits the lake only on migration. Occasionally, it may be seen over the pastures, giving the watcher an opportunity to observe its guerrilla tactics. It dashes low over open terrain to hunt ground squirrels or pay a surprise visit on a party of magpies in willow shrubbery. Ducks react fearfully whenever the big accipiter comes near. Soaring over the wooded east shore, goshawks have been seen to dive after lower flying mallards or pintails, but when the hawk is serious about hunting waterfowl, it stays low and darts among the reed beds, scattering ducks and shorebirds like leaves in a storm. Coots are frequent victims that may be carried back to the woods to be plucked and eaten in a secluded corner.

Cooper's and sharp-shinned hawks nest here and there in the densest woodlots around the lake. Because of their partiality to hunting under cover, their depredations are seldom observed. But during late summer, these secretive hawks do some of their foraging in the open, giving us a chance to compare their style to that of the falcons. Instead of approaching their prospective victims from afar, the cooper's and sharp-shin will sneak up, pausing in low bushes or on the ground. The final sprint is very fast, but if the threatened birds rise in time the chance for success is half-lost for the hawk.

An unfamiliar hunting method of the sharp-shin is a falcon-like stoop, usually performed during migration when the small accipiters often soar very high. They are fond of attacking feeding snow buntings, falling like a meteor out of the blue, scattering the startled birds in a roar of small wings. In one successful hunt, the sharp-shin ate its kill among the weeds of an open field, far from cover.

Like falcons, accipiters use obstacles, natural or man-made, to conceal their approach in a deliberate attempt at surprising the quarry. Individual hawks will develop a trick or two of their own, and adults will be more efficient and versatile than immatures. Young accipiters are fond of harassing birds too large for them to kill.

The forest hawks and the ''long-wings'' appear to have a hearty dislike for each other. The goshawk may kill and eat any falcon it can catch, but it treats the smaller accipiters in the same manner. The peregrine is no better and will readily stoop to kill any sharp-shin or kestrel that flies below it. One day, a peregrine was seen in deadly serious pursuit of a merlin, a unique opportunity to compare the flying powers of these two champions. The bigger falcon outflew and outclimbed the smaller one, but only just, and the nimble merlin managed to escape high into the sky. Falcons are generally believed to be ''nobler'' than the rapacious and sneaky accipiters, but in practice their motto is the same; grab what you can, where you can, and how you can.

Chapter 11

CRANES, RAILS AND COOTS

The order of Gruiformes includes an incongruous collection of species that diverge in size from the stately whooping crane, standing almost as tall as a man, to the mouse-like yellow rail that skulks unseen in the wet grass. No doubt, there are structural and physiological similarities that hold this group of birds together, but on the surface they seem to have little in common except a preference for marshy habitats.

The great whooping crane is believed to have been common on the prairies before settlement, perhaps nesting at Beaverhills Lake, before the conspicuous giant was brought to near-extinction on the continent. The breeding range of its close cousin, the sandhill crane, formerly included suitable habitat far south of the boreal forest, but it is still abundant in its present northern range, as evidenced by the massive migration that take place across Alberta each year. The spring passage of sandhill cranes is one of the great events on the birdwatcher's calendar. If one is lucky enough to be afield on a peak day, the sky over the lake is full of birds as far as the binoculars can see. Estimates exceed ten thousand cranes in a single morning, which makes one wonder about the total size of the North American population, considering that migration takes place across Canada's interior from the Rocky Mountains to Hudson's Bay.

With its long neck and spindly legs, the sandhill crane resembles the great blue heron, but the two differ as much in habits as in character. A solitary fisherman or lone traveller, the heron may give vent to the occasional screech if it is disturbed, but the social crane produces one of the most delightful sounds in the avian repertoire. Melodious and resonant, the rolling calls drift down from the sky long before the birds are visible to the naked eye. Scanning through binoculars in the direction of the sound, we may spot the birds still more than two kilometres away, soaring high among the clouds, or winging in line formations over the distant horizon.

The great carrying capacity of the crane's voice is due to the length of the windpipe that stretches over one metre from the bird's throat to the lungs. During flight, the cranes call to keep contact with each other, so that the vanguard knows where the stragglers are, which is especially important when the birds travel at night. Perhaps they also call out of a sense of contentment, like a purring cat. The frequency of their notes increases when the flock prepares for landing, as if the birds need to check and double-check their position while examining the ground below. Buoyant when strong winds are blowing, the birds descent slowly and very steeply, their wings held up like an umbrella. The legs are lowered well before reaching the ground. At any moment, the intended touch-down may be aborted or postponed, with the birds wheeling around, all the while keeping up their carolling conversation.

Once the flock has landed, the birds fall silent, looking about warily before relaxing. Occasionally, a pair may suddenly perform a little dance with wings spread, as if overcome with joy, a nuptial ballet that will be fully displayed once the birds reach their breeding grounds on some wilderness muskeg.

Crane stop-overs at the lake are brief and occur mostly in late evening. By morning, the flock will resume travel, but a few birds may linger a day or two. They walk quietly on the fields, searching for grain, insects, frogs, or the occasional vole. On the pastures, cranes have a habit of pecking apart cow patties to get at the beetles and flies underneath. No farmer would object to such manure spreading services performed by the cranes, but the big birds can do some damage on unharvested grain during their fall visits. Passing high over the lake in August and September, the migrating flocks traditionally descend and congregate on the wide open plains farther east, especially in Saskatchewan and Manitoba, where hunters are encouraged to prevent crop depredations. Apparently, the meat of cranes is quite palatable, in contrast to that of the fish-eating heron with which the crane is often confused by the shooting fraternity. Hunting of cranes is not permitted in Alberta, at least not at the present.

A shooting season remains in effect on an unlikely relative of the crane and fellow-member of the Gruiformes, the American coot. Weak on the wing, the ''mud hen'' keeps its distance from people by swimming away with nodding head, or, if in a hurry, by skittering along the surface with fluttering wings and pattering feet. However, once airborne, the coot's stamina is apparently sufficient to carry it to California on migration. It returns to central Alberta early in April, travelling under cover of darkness and announcing its arrival by its ''hammer-hits-stake'' calls, while it remains hidden from view among the reeds.

Deceptively timid and retiring, the coot is pugnacious and violent in disputes over nesting territory with its neighbours. Like a pair of mud wrestlers, lying on their back, the combatants thrash each other with their big, lobed feet. In water they seem intent on drowning the opponent, pecking its head in a frenzy of aggression. The victorious bird, its

ego inflated, swims about with wings fluffed over its back like a tent. From behind, the two white spots in the raised tail look like the eyes of a madman.

Coots raise large families. They often lay more than a dozen eggs in a cozy nest built from plant material. The young look like little black devils with a bald head, red bill and bristling, bright orange down around the neck. Omnivorous, consuming vegetable matter as well as insects and crustaceans, the chicks find plenty to eat and enough survive to replace the adults that fall victim to hawks, eagles and owls, or to the duck hunter in fall and winter.

It seems odd that a shooting season is also in effect on rails, although no-one seems to be hunting them for the simple reason that these secretive little birds remain hidden among the reeds. Moreover, compared to a well-fleshed mallard or even a coot, a rail can hardly be considered a worthwhile bag. ''Thin as a rail'' is proverbial, but this adage has nothing to do with the birds's lack of fat, and refers instead to its narrow, laterally flattened body shape, an adaptation to its habitat. The slim rail easily slips through narrow spaces between the reeds, gaining a foothold on soft ground on oversized feet with long flexible toes. Gingerly stepping on floating duckweed and aquatic plants, a rail can practically walk on water.

Three of the half dozen rail species occurring in Canada have been identified at the lake: sora, yellow rail and virginia rail, although the latter is locally very rare. The sora is common, even numerous, as evidenced by its calls emanating from wet places everywhere. It is an easily recognizable whinny, a series of sharp notes on a descending scale, given in a strident, garrulous tone. An imitation of this call, or any scratchy noise, may attract the rail out of curiosity or annoyance. If we stand still near the edge of the reeds, a sora may suddenly emerge a few metres away. It is a thrill to finally see this pretty bird so close!

Rails seldom take flight. And if they do, we get only the briefest glimpse of their rounded wings and dangling feet before the weak flyer drops back into cover.

Considering the vast differences between the lofty sandhill crane and the tiny sora rail that skulks in the marsh, it seems hard to believe that the two species are related and belong to the same avian order.

An even more elusive rail, the yellow rail, was formerly thought absent or extremely rare at the lake, until quite recently. It too has a distinctive call, an oft-repeated ''tik, tik, tiktiktik'', as if two stones are clicked together in a monotonous rhythm of two slow and three quick notes. The yellow rail's presence was first reported by the late Loran Goulden, who camped at the lake and recognized the diagnostic night-time noise from a tape recording of marsh birds made elsewhere. Fellow birdwatchers, alerted by Loran's report, have since heard the yellow rail each spring, especially in the evening and during the night, but also in the daytime.

However, *seeing* a yellow rail is another matter and requires a little strategy and patience. The calling bird, hidden in some wet grass or rushes, should be approached slowly, step by step. The clicking call is surprisingly loud when heard close by, at about twenty metres. Further advance will cause the rail to fall silent, and it won't start again for a long time, even if we stand perfectly still. A quick walk into the direction of the last clicks may flush the bird out of hiding and scare it into flight. In good light, there may just be time to see its yellow and brown plumage before this robin-sized recluse descends into the rushes again.

The Gruiformes include some diverse characters indeed, from the conspicuous sandhill crane, a delight to the eye and the ear, to a tiny rail that allows the average birdwatcher little more than a one-second glimpse in a lifetime.

Chapter 12

SHOREBIRDS - THE SPRING MIGRATION

Easy to find in their open environment, shorebirds are a delight to watch because of their lack of shyness and the great variety of shapes, sizes and colours. Not counting a dozen or so vagrants and rare visitors from other continents, North American shorebirds include about fifty species, an amazingly large number, considering that they are crowded together on a narrow strip of habitat between water and land where they must share similar food resources.

To lessen competition, the various species have evolved divergent physical features, such as length of legs and bill, that permit a division of habitat through spatial segregation. The squat dunlin hunts for insects and crustaceans along the waterline, where it does not run into the stilt sandpiper that feeds in deeper water. Godwits and avocets, by virtue of larger size and longer legs, can wade even farther out, until swimming shorebirds such as phalaropes have the surface to themselves.

Ecological segregation and a sharing of food resources is further achieved through divergence of bill structure and feeding habits. Dowitchers and snipe, with their long and sensitive mouth pieces, probe deep into the mud for invertebrates not available to the needle-billed yellowlegs or a stubby-beaked plover. Semipalmated and least sandpipers are nearly identical in size, but the whitish semipalm feeds habitually in the shallows, whereas the brown-backed least has a preference for mudflats. The slightly larger pectoral sandpiper often forages in wet grass, away from its two smaller relatives. Of course,

at some time or another, all three species may find themselves together on the same piece of turf, but if they meet, their intolerant and aggressive attitude, towards each other as well as their own kind, serves to separate the individuals.

Spatial segregation is very evident in golden and black-bellied plovers, that join several other species on the lake shore for resting but fly inland to forage on meadows and fields, away from the competition. Most plovers and some sandpipers are actually dryland birds that inhabit upland prairies or arctic and alpine tundras during the breeding season, and arid savannahs while wintering in South America. Considering this diversity in habitat preference among the species, the collective term "shorebirds", used exclusively in North America, is not entirely correct, but neither is the British equivalent "waders", since some plovers, sandpipers and turnstones seldom stand in water deep enough to wet their feet.

During migration, the various kinds of northern shorebirds travel by a staggered timetable which further lowers competition during feeding stop-overs along the way. Differentiation in the timing of migrants is very pronounced at the lake between the end of April and late May. The spring passage of the first shorebirds, mainly yellowlegs and the odd hudsonian godwit, subsides when the next wave, consisting of dowitchers and pectoral sandpipers, arrives. Smaller sandpipers, colloquially called "peeps" come later. Of course, there is a certain amount of overlap, but the last wave, made up of black-bellied plover, red knot, sanderling and buff-breasted sandpiper, come latest and stay longest.

The lake's reputation as a stop-over for arctic-bound shorebirds is largely based on the glowing accounts of the late Professor William Rowan, who studied and collected birds during the first half of this century. But many a modern-day visitor has gone home disappointed, wondering whether the lake's fame is exaggerated or whether there has been, for some insidious reason, a disastrous decline in its birdlife. This may well be true, but hopefully only on a short-term basis.

Of course, some years are better than others. Although migrating birds travel by a timetable that is as predictable as the seasons, where and when they will touch down along the way, or in what numbers, is as capricious as the weather. This is especially true for sandpipers and plovers, those restless globe-trotting wanderers that make twice-yearly flights between our arctic coasts and South America. In some springs, northern shorebirds of a dozen species build up at the lake to astonishing numbers. In other years, the flocks barely exceed a few dozen birds, here today, gone the next.

Based on the author's extensive field notes covering 25 years from 1964 to 1989, spectacular spring migrations, such as those described by Rowan, were the exception. In most years, bird numbers were low or moderate. Especially during most of the 1980s, if a build-up happened at all, it was very local with one or two thousand sandpipers on a favourite pasture and very few on miles of adjacent shoreline. Unless the birdwatcher is prepared to search all of the lake, not just the popular access areas, it is easy to miss the birding hotspots of the season.

As discussed in other chapters, the reason why much of the lake is shunned by wading birds has to do with habitat quality. Wide margins of cattails and bulrush have overgrown formerly open shallows, making them unattractive for sandpipers and plovers. A few sections of shore that have remained open and free of reeds, such as the wave-swept south shore, have a sandy bottom that offers limited food resources. Remnants of superior muddy shallows exist along some overgrazed pastures along the west shore, where cattle, year after year, trampled the soft ground, destroying emergent vegetation and preventing reeds from growing.

Spectacular shorebird spring movements, such as those of 1977, 1978 and 1989, occurred during periods of drought and dropping water levels, and coincided with a high nutrient cycle of the lake, creating habitat as well as abundant food, the twin elements required for the prosperity of all life on this planet.

Good shorebird years are not necessarily preceded by an early break-up of the lake ice. The arid spring of 1988 was particularly poor for shorebirds despite the record date on which the lake was free of ice, April 17, a month ahead of late years such as 1967, 1979 and 1982, when part of the lake was still frozen on May 17. However, spring temperatures play a decisive role. If the lake water is still cold by the time northern shorebirds arrive, they are forced to move since food is scarce. On the other hand, if an early hot spell has triggered the hatch of chironomid midges, locally called lake-flies, shorebirds linger for days or even weeks, rebuilding fat stores before resuming their journey to the arctic.

In springs of average temperatures, the lake-fly hatch begins around the middle of May. Warm weather advances it, cool springs delay it until June. If the hatch is massive and coincides with the arrival of the first sandpipers, the stage is set for an exceptional build-up. In 1977 and 1978, lake-flies were already swarming by May 8. On calm evenings, the insects hovered over the shore like smoke, visible from miles away. Windy conditions forced the flies down until they plastered bushes like a black mould. Where they crawled about on the grass, shorebirds collected to gorge themselves. On May 19, 1978, buff-breasted sandpipers were estimated at 1,500 along the north and west shores. Pectoral sandpipers exceeded 10,000. On a

Dull grey and quite rare in fall, red knots in spring plumage are common at the lake in late May and often associate with black-bellied plover. Buff-breasted sandpipers (top) are also far less regular in fall than in May, when they may interrupt their feeding to engage in a silent ritual. Raising one or both wings, they "wave" at each other in mating display.

point along the east shore, there were 300 black-bellied plovers, 200 red knots, 150 sanderlings and dozens of ruddy turnstones. Mixed flocks of pectoral, semipalmated, least and Baird's sandpipers were scattered all along the lake. On June 4, 1978, white-rumped sandpipers were recorded in unprecedented flocks, totalling an estimated 1,100 birds. A hundred or so were still present on June 17, an exceptionally large number for such a late date.

As stated before, other crucial factors determining the arrival and numbers of migrant shorebirds are weather patterns which in turn depend on the location of the jetstream. High pressure cells and persistent westerly winds are associated with poor birdwatching at the lake, whereas exciting days are heralded by southeasterly winds that precede a low pressure system and bring in migrants that otherwise might pass over neighbouring provinces. While it is known that shorebirds travel north from the Gulf of Mexico across the continent with few stops, very little or nothing has been learned about the route. Is it always straight or sometimes curved and deflected by weather systems? During the very dry spring of 1988, buff-breasted and stilt sandpipers, which are seen each year in good numbers, were scarce. Assuming that the birds did make their journey in 1988, the main body of these species probably followed a route that did not pass over central Alberta. In 1989, which was an excellent year for spring migration of shorebirds, buff-breasted sandpipers were not recorded at all at the lake. Interestingly, reports from Saskatchewan localities indicate that shorebird migrations were spectacular in the spring of 1988 and poor in 1989, exactly the opposite from Beaverhills Lake!

In the future, perhaps through the imaginative use of miniature satellite tracking devices, researchers may be able to follow individual birds across the globe, which would greatly improve our understanding of the mysterious forces that compel a tiny puff of feathers like the white-rumped sandpiper to travel twice-yearly between northern Canada and land's end at Tierra Del Fuego, a distace of 15,000 km!

A few sandpipers (top) or dowitchers (bottom) may arrive before the end of April, but the main spring migration takes place after the first week of May and peaks around the 20th of the month.

Shorebirds such as the marbled godwit lay relatively large eggs to give their young a head-start in life. To take up as little space as possible under the brooding bird, the eggs are pear-shaped and fit together like wedges in a pie.

Chapter 13

SHOREBIRDS - THE LOCAL BREEDERS

The spring arrival date of locally breeding shorebirds, namely killdeer, marbled godwit, willet, avocet and Wilson's phalarope, varies much more than the date for arctic migrants that come from farther away and have farther to go. The earliest of the locals, the hardy killdeer, winters not far from Alberta, and the odd individual may venture north during a mild spell in mid March, weeks ahead of all other shorebirds.

A typical dryland plover, the killdeer does not seem to care whether the lake is still frozen and the country snow-covered as long as there are a few patches of bare soil. The first killdeer is usually heard before it is seen, flushing ahead of the watcher with the familiar cheery call. In flight, it shows the rufous colour of the rump that is hidden from view when the wings are folded.

The diagnostic double breast-band of the killdeer is an example of so-called disruptive colouring that breaks up the bird's silhouette in the interest of camouflage, at least as long as the bird remains immobile. Disruptive colouring by way of contrasting patterns is also prominent in the black-bellied and golden plovers, as well as in ruddy turnstone and dunlin, but in these species it is a feature of the breeding plumage only, restricted to the time when these birds inhabit tundras patterned with lichen-encrusted stones and bright flowers.

The local killdeer nests on bare ground or very short grass, along the shore as well as inland, but suitable terrain is always at a premium. Often, the bird has to resort to roadsides or parking lots. Around the lake, killdeers have taken to nesting on the gravelled gas company roads that disect the pastures. The birds do not make any attempt at nest building, but lay their four spotted eggs in a slight depression, occasionally decorated with a few stones or twigs. Well before a person gets near, the brooding bird slips off the eggs quietly and runs for a distance until it bursts out in vehement protest. Especially later in the season, when the precocious young have fledged, the alarm calls of killdeer accompany the hiker all along the lake's south shore.

Of course, it is prudent to stay away from the centre of the bird's concern, since the eggs or the crouching young can be stepped upon unwittingly. However, before the risk becomes imminent, the killdeer changes its strategy and attempts to lure the intruder away by so-called distraction behaviour. The bird drags itself along the ground as if injured and unable to fly, while it screams pitifully. No doubt, such tricks have proved their value to distract coyotes or foxes, unless they too get wise and refuse to follow the injury-feigning bird, that jumps into flight as soon as the predator gets too close for comfort. Ravens have been seen to search for and find a killdeer nest while ignoring the protest calls and distraction displays from the desperate owners. There is no reason to believe that other corvids such as crows and magpies are less smart than the raven, but they are smaller and may respond more readily to the killdeer's belligerent demands "to get lost". No amount of scolding or distraction would help if the nest was approached by a cow. In frantic and effective attempts to turn the innocent but dim-witted bovine around, the killdeer literally "flies in the face" of danger.

Like killdeer, avocets nest on bare ground and the brooding bird slips away well ahead of approaching people. It joins other members of the loose colony in collective protest. By contrast, brooding shorebirds of species that hide their nest in grass, such as godwits and willets, stay put and leave the territorial defense to the non-brooding partner, that flies out to meet the intruder in an agitated and noisy manner. Willets swoop aggressively over a person's head in a rush of wings. But the brooding bird will flush only if you stumble on the nest. Some godwits or willets sit so tight that they can be touched. They have even allowed themselves to be picked up by people for a brief look at the eggs!

Godwit eggs, like all shorebird eggs, are relatively large and beautifully peryform or pear-shaped. Four of them, the usual number in a clutch, fit together like wedges in a pie, taking up the smallest possible space so as to fit under the brooding bird's body. Unlike ducks, both partners of a mated pair of shorebirds share incubation, which amounts to a more efficient use of individual energy resources. The eggs need a fairly brief incubation period of 17 to 24 days, and the young are well-developed. Within hours of hatching, they are quick on their feet and capable of catching insects. Mother Nature has given them a headstart in reaching the flying stage, and as soon as they are fit to travel, the parent birds will accompany them on the flight south. The young of some arctic sandpipers are ready to fly in 14 days.

To speed up the breeding cycle and to share the risks of reproductive duties even more fairly among the sexes, phalaropes have reversed the sex roles in the care of eggs and young. It is the male that does the brooding, while the female spends her time loafing and eating, soon after she has laid the eggs. This unusual arrangement ensures that the female recuperates quickly from the energy loss associated with the production of her large eggs.

The male Wilson's phalarope is a charming little fellow, and quite unlike other locally nesting shorebirds, he is not noisy. When flushed from the nest, hidden in the grass, the male flutters away, uttering low grunts. He may be joined by other males that nest in the vicinity. In the early part of the breeding season, the females assist their mates in the protest demonstration, flying about in a loose flock. The brightly coloured females are quite aggressive in courting the duller-looking males, but once incubation is in progress, the ladies lose interest in their partners. By the end of June nearly all females have left central Alberta and are on their way to winter quarters.

The raising of young phalaropes is entirely up to the males, which depart soon after their brood can take care of itself. By the end of July, it is rare to see any Wilson's phalaropes at the lake. By that time, also the calls of godwit, willet and avocet have fallen silent, and only a few stragglers will be around in August or early September. However, the odd killdeer hangs on until freeze-up in late October or early November, but its vociferous vigilance towards people is spent for another year.

Folding its long legs under its body, an avocet settles down on the eggs. It nests at the lake in loose colonies along some muddy sections. Wading or often swimming, it harvests insects and crustaceans with scythe-like sweeps of its upturned bill.

Deceptively docile and nondescript when at rest, the willet will spread its black-and-white wings and jump to the defense of its youngsters well before a person comes near. Screaming, it will swoop at the intruder's head until he or she has left the neighbourhood.

In phalaropes it is the male that broods the eggs and raises the young. This unusual arrangement allows the female to recuperate quickly from the energy loss associated with egg laying. In the picture, the female tries to crowd the male off the nest so that she can lay an additional egg. Dutybound, he seems loath to leave...

The double neck-band of the killdeer (opposite page) is an example of so-called disruptive colouring that breaks up body shape in the interest of camouflage. The plumage of killdeer chicks is also highly cryptic. The baby bird all but vanishes from sight among the ground litter as long as it holds still.

The common snipe keeps a low profile in the marsh, probing the soft ground with its specialized bill that has an extra hinge at the tip of the mandibles to grasp a worm deep in the mud. During the mating season, the timid snipe may be bold enough to perch on a post. In spring, it flies high over its territory in a peculiar zigzagging flight. On the downward rush, its vibrating tail feathers produce a winnowing sound that some local farmers associate with the coming of rain.

Chapter 14

SHOREBIRDS - THE FALL MIGRATION

After the passage of spring migrants has stopped, there is usually a brief period from mid June to early July when a visit to the lake fails to turn up any shorebirds other than the five species that nest locally. Return migrations from the arctic seldom begin until July, but in some years, perhaps when severe weather conditions in the north disrupt breeding, there may be a certain amount of overlap between spring and fall passage. Of course, it is not possible to know whether a particular shorebird seen in late June is a straggler on its way north or an early returner going south. For instance, on June 24, 1975, there were two pectoral sandpipers, two stilt sandpipers, two red knots and six hudsonian godwits along the north shore. Were these birds on the way up or down the continent?

Also noted on that date were half a dozen dowitchers and lesser yellowlegs, species that nest not far away in northern and western Alberta, but are usually absent from the lake between late May and early July. Sightings of these birds in the last week of June probably indicate a return from the breeding grounds. Thus, while one species is still moving north, another may be coming back, albeit from lower latitudes. (It should be noted that lesser yellowlegs were reported nesting in the Tofield area in the first quarter of this century, but the author has seen no nests at the lake since 1965, although one pair displayed alarm behaviour as if it had eggs or young in a wet, grassy area along the south shore.)

By early July, the return migration of arctic shorebirds is definitively on its way. For instance, on July 5, 1975, there were several flocks of peeps, including least, pectoral and Baird's. And one week later, there were fifty stilt sandpipers and some semipalmated plover. After the middle of July, the number of shorebirds increases rapidly. During the two weeks preceding July 26, 1976, the species total, including the local breeders, was 26, a remarkable tally for mid-summer when few birdwatchers visit the lake or expect to see much there.

Like the spring migration, the fall movement and build-up of shorebirds varies yearly, with the most exceptional numbers recorded between 1975 and 1978, the period when the lake receded from the high waters of 1974. The decomposition of flooded shoreline vegetation created a nutrient-rich environment for aquatic insects and crustaceans, which attracted the birds.

The long time-span of fall migration, from July to October, and the low number of shorebirds compared to spring, allows the watcher to pay close attention to the identification of individual birds, especially of the confusing sandpipers. We should attempt to get near without flushing them. Sitting on a stone along the shore, with the sun in our backs, we can observe the sandpipers at leisure through binoculars or telescope. Eating on the run, the peeps approach closely and newly arriving flocks may land nearby. With patience, you will get excellent views from less than ten metres. It is a thrill to have the lively birds so close, ignoring or tolerating our presence.

Detailed observation and comparison of species seen together is the best way of

The number of golden plovers that make a migration stop-over at the lake seems to have declined over the years. In fall, when the plover is a subdued shade of speckled brown, only the odd one will show up along the shore.

becoming familiar with the four commonest sandpipers: least, semipalmated, Baird's and pectoral. All of them have the same general colouration, counter-shaded with light bellies and darker backs, which serves to dissolve the tell-tale shadow for better camouflage.

Instead of looking for just one diagnostic feature, we should note all of a sandpiper's characteristics. For instance, apart from its yellowish legs, the least sandpiper has a slightly curved and pointed bill, and its back is darker than in the semipalmated, which has black legs and a straight, rather blunt bill.

A confusing peep is the Baird's sandpiper. It has black legs and a straight bill, like the semipalmated, but its plumage is more buffy in colour with scaly markings on the back and a more densely spotted breast, sharply defined from the whitish belly. When seen together, the Baird's is slightly larger and slimmer than the least or semipalmated, but it closely resembles the white-rumped sandpiper in shape. However, when it flies up, the Baird's does not show a white patch on the base of the tail, which is diagnostic for the white-rumped.

Fortunately, one of the commonest sand-pipers, the pectoral, is easy to recognize on account of its yellow legs and streaked breast shield. Its bill is rather long and slightly curved. When flushing, the pectoral utters a distinctive, grating alarm call.

Once the watcher has mastered the skill of distinguishing all of the above peeps, a skill that may have to be relearned or honed each fall, one can confidently identify the rarities, such as the western and sharp-tailed sand-pipers. Both have been reported about a dozen times in the past twenty years, but they may be less scarce than the records indicate. On September 27, 1987, the author saw the two celebrities together, under very good light conditions and from no more than ten metres away. The western sandpiper's legs were black, like a semipalmated, but its bill was slightly curved like the least and perhaps a touch longer. The clinch was the rusty streak on its shoulders, which is a diagnostic feature of the western, a species that is common only along the Pacific coast and seldom strays inland.

Sharp-tailed sandpipers might have been overlooked at the lake or dismissed as off-colour pectorals, which they resemble, were it not for the vigilance of a trio of Edmonton birders who first identified the species in October of 1975. Their slides, shown at a meeting of the Edmonton Bird Club, alerted other members and sparked additional records in 1978, 1979 and 1980. Up to four birds stayed around for several weeks between late September and early November. Since then, only the odd single sharp-tailed sandpiper was seen in 1985-1990. It can best be identified by the unmarked breast, which is light buffy in adults and a deeper cinnamon buff in first-year birds. Also the reddish-brown cap and the prominent white eyebrow are good features in the field that distinguish this Siberian vagrant from the common pectoral sandpiper.

Several species of shorebirds are much harder to identify in fall than during spring and summer. After losing the chromatic exuberance of their nuptial plumage, dunlin, red knot and red phalarope are dull grey birds that can only be recognized by body shape, bill length and feeding habits. Each of these circumpolar wanderers migrates mainly along the coasts and rarely shows up far inland during fall. Even common species such as black-bellied and golden plovers, sanderling and spotted sandpiper, look plain dull in their winter plumage and may be difficult to name for the novice birder.

Quite apart from the opportunities for mere identification, the long fall days provide a chance for studying shorebird feeding behaviour. Throughout the year, these intense, hyperactive creatures spend much time foraging on sea coasts where their intertidal prey is only available for part of the day during low tide. The birds are therefore under pressure to forage efficiently. In contrast to dabbling ducks, which scoop up their food indiscriminately, sieving crustaceans and plant seeds through their specialized bill, shorebirds select food items one by one. Recent studies have shown that the various species do not necessarily concentrate their efforts on the most common or the largest prey, but on the most profitable one that yields maximum returns. Sensitive to potential feast or famine, shorebirds congregate in areas of highest food densities and each species develops a so-called searching image for their choicest prey, switching to alternate kinds if the former becomes harder to catch or grows too large to handle. These skills are evidently perfected or learned by experience, since adult birds were found to be more efficient at foraging than juveniles.

Among shorebirds, there is a clear difference between hunters and harvesters. The former rely on sight to locate prey, the latter use touch. Typical for a "hunter", the yellowlegs paces the shallows, searching for scud, tadpoles and insect larvae stirred up from the bottom. Plovers hunt the shoreline in a deliberate stop-and-go manner, looking intently for prey while standing still and dashing a few steps ahead to snap it up from the ground. One day, a killdeer caught a three-centimetre-long water beetle, which was hammered and shaken for twenty minutes before it could be swallowed. Despite the absence of claws or teeth, the killdeer is a predator in its own right. It is particularly fond of beetles and grasshoppers. The larger shorebirds, such as godwits and willets, occasionally capture small fish that are shaken and pinched before swallowed.

In contrast to the deliberate stalking and stabbing behaviour of the "hunters", the avocet gathers its prey haphazardly, sweeping the shallows with scythe-like motion of the curved bill and seizing food items that happen to be hit. Other harvesters, such as dowitchers and snipe, probe the mud with their long bills, which are equipped with an extra hinge near the tip of the mandible that allows the bird to grasp a worm deep in the soft ground. Interestingly, touch feeders like dowitchers do not mind working in a dense group, since their prey are not aware of the enemy. On the other hand, hunters are intolerant of crowding that would reduce individual success. Hence, yellowlegs and plovers space themselves out during foraging, whereas dowitchers stay together in a compact flock.

The singular ruddy turnstone is in a class by itself. It is a hunter as well as a scavenger that habitually turns over shoreline wrack and consumes anything edible, live prey and carrion as well as vegetable matter such as berries. It has also been seen to eat bread and the eggs of other birds.

As the summer wanes and the waters cool, food becomes less abundant for shorebirds. Yet, up to a dozen species may hang on around the lake to the end of October, and the odd bird lingers until winter begins in earnest. Black-bellied plover are among the last to leave. Their fluted calls take on a wistful quality as they fly up from the frozen mud. It is with sadness that we see them go, and we look forward to their return in spring.

Photo: Dick Dekker

Photo: Dick Dekker

Photo: Richard Klauke

Some of the rarer shorebirds that have been recorded at the lake.

1. The sharp-tailed sandpiper differs from the pectoral by the unmarked breast, dark cap and prominent eyebrow.

2. In winter plumage, the red phalarope can be told from the more common red-necked phalarope by the absence of streaks on the back.
In addition to one spring record of a female in full breeding plumage, in May 1970, there are at least half a dozen sightings of up to four birds in late September and October.

3. This surfbird was seen by a group of Edmonton Bird Club members who happened by when the photographer had just discovered the Pacific wanderer on the south shore of the lake.

Western sandpipers have been identified about a dozen times during September. Six were captured and banded in May 1990.

Chapter 15

GULLS AND TERNS

Raucous and uninhibited, gulls are conspicuous at every human activity that offers something to scavenge. There is no magic about "seagulls", none of the elusive quality we associate with wild things and wild places. Few birdwatchers would rate them as their favourite species. Yet, gulls are admired for their mastery of unruly air over stormy seas, and their lack of shyness makes them easy and entertaining to watch.

Glaringly white and visible from afar, gulls seem to spurn mimicry or cryptic colouration, the anti-predator device used by all hunted animals to lessen the danger of discovery by their hunters. Although juvenile gulls are brownish grey to help them escape predation on the nesting grounds, once they have reached maturity, adult gulls have few enemies. Even the redoubtable peregrine falcon seldom bothers to attack the shifty gulls. As hunters of fish, gulls are predators in their own right. And it is to hide from the eyes of fish that gulls have evolved white undersides. Hovering over the waves and ready to plunge, the light-coloured bird blends with the sky.

Curiously, some of the smaller gulls, such as the Franklin's and the Bonaparte's, both common at the lake, have black heads, but only during summer when they eat mainly insects. By the time the gulls change back to their winter diet of fish, the black feathers are moulted out and replaced by white. Why do these gulls bother to acquire a black head in the first place? Apparently, according to the famous Dutch ethologist Niko Tinbergen who studied colonies of black-headed gulls in Britain, the mask is used in behavioural rituals during the mating and nesting period. The black head in combination with the red bill and the white-accented eyes scares off the neighbours and keeps them from nesting too close-by. Gluttonous and voracious, the gulls would not hesitate to devour a neighbour's unguarded eggs or tiny young. An additional reason why colony-nesting gulls need to space their nests well apart is to prevent mass losses to predators and scavengers. If the nests were very close together, a raider would have little difficulty in finding them all. But spaced several metres apart, even a sharp-nosed fox will locate only one set of eggs at a time and miss most of the others.

Why do gulls bother to nest in a colony when there are such serious risks involved? Because colony nesting has some distinct advantages too. Together, the birds can effectively defend their nests against common enemies such as large gulls, crows and hawks. Moreover, the occupants of the colony benefit from the communication of foraging opportunities. The eager-eyed gulls watch each other jealously, as anyone knows who has thrown bread to the birds at a park. Soon after one gull discovers the bonanza, others hurry in from afar to fight for their share.

It appears then, that the gregarious gulls are caught between conflicting impulses; the urge to seek each other's company for collective and individual benefit, and the need to stay apart for the safety of their brood. A balanced arrangement is worked out by constant

One of the earliest migrants to return in spring, the ring-billed gull (top) is common until freeze-up. The pretty sabine's gull (bottom), that breeds on arctic coasts, is one of half a dozen very rare gulls that have been recorded at the lake.

bickering and aggressive posturing. A gull colony surely is a noisy place!

A good question is: how do black-headed gulls attract a mating partner if their facial mask is designed to keep conspecifics at a distance? The gulls prevent unnecessary friction by avoiding direct stares. When a male and female meet, courtship on their minds, the birds assume an upright posture, craning their necks, and, as if on signal, both turn their heads and look away. They hold this non-aggressive pose for a second before glancing back to observe the other's reaction. Once mated, the pair stands side by side, each facing into the opposite direction, avoiding each other's black mask.

Franklin's gulls breed in several large colonies in the reed beds off the lake's northeast shore. The number of pairs was estimated at 10,000 to 20,000 in 1976, but they are far less numerous now. The nests are placed on mats of dead vegetation in between cattails and bulrush.

Bonaparte's gulls do not breed at the lake, although they are numerous in spring and late summer. These pretty little gulls inhabit muskegs in the boreal forest, where they select a nest site in a spruce tree, well above ground. Probably, these gulls have adopted this lofty habit because of a general scarcity of suitable islands with low vegetation.

At the lake, open island habitat is at a premium too. The only suitable site is taken up by the larger gulls, ring-billed and Californias, that have to compete for space with pelicans and cormorants. Other island-nesting birds, such as common terns and avocets, have to make do with gravel bars here and there. Of course, avocets also inhabit mudflat habitat along the shore, but terns and gulls avoid the mainland and insist on islands that guarantee a certain amount of protection from mammalian nest robbers such as coyotes.

Like all colony-nesting birds and many solitary species, gulls are very traditional in their choice of nesting site, returning year after year. A crisis occurred in 1974 when the lake rose about one metre and inundated the

breeding islands. The gulls began nesting on the mainland and on points temporarily cut off from the shore by the rising waters, until dropping lake levels forced them to move again. Some new settlements failed completely after the eggs disappeared. By 1976, the gulls could return to one of their pre-flooding nest sites off the east shore. This island was later raised with rocks by the Fish and Wildlife Division. Named Pelican Island, it was declared a Natural Area in 1987. During the breeding season, Pelican Island is out of bounds for people, which is an excellent idea because visitors can raise havoc with any colony by forcing the birds off the nest, risking the survival of small chicks during cold or hot weather.

A visit to a gull colony is not an unmitigated pleasure. The birds do not hide their annoyance and fly overhead in shrill protest, hitting the visitor, purposely or accidently, with squirts of excrement. During a visit in 1977, many dead young were found. Dead adults had ground squirrels protruding from their gullets. The cause of death of these gulls was probably strychnine from poisoned squirrels picked up on fields near the lake.

The large California and ring-billed gulls are true omnivores, but their favourite food is fish. In early spring, they are among the first birds to return to the lake, congregating at the mouth of inflowing creeks when sticklebacks and minnows travel upstream to spawn. After these small fish become scarce, the gulls fly inland, perhaps to forage at other lakes. They also visit fields where voles are abundant, and they never pass up a chance to steal eggs of ducks and shorebirds, or even those of short-eared owls.

Later during spring, some of the large gulls become persistent hunters of ducklings. They fly low over the reed beds and swoop down to grab a baby duck before its mother has a chance to intervene. If she has time to gather her brood, a female mallard can put up an effective defense by jumping up at the attacking gull, wings thrashing the water. The gull will try again and again, until it gives up

The elegant black tern is abundant on shallow water all over Alberta. Arriving in May, it will be gone again by early August. Foraging mostly on insects, it may treat its fledged young to a few small fish.

to look for a less troublesome opportunity elsewhere. If successful, the gull seizes the victim in its bill and, without bothering to alight, carries the prey aloft and swallows it along the way. These little tragedies probably happen often but are seldom witnessed by people. Few gull-watchers visit the lake during summer, and those who do concentrate their attention on spotting the rarer species.

The list of vagrant and transient gulls from more northerly regions that have been identified at the lake is quite impressive. Among the larger species are herring and glaucous gulls. Especially the brownish immatures, lacking the diagnostic features of the adults, are hard to distinguish. Rare smaller gulls that have been reported include the decorative sabine's gull, the mew gull and the black-legged kittiwake. In 1987, a keen birder spotted the first little gull, an elusive species that is reported increasingly often in Canada after it was first discovered nesting in Ontario in 1962. Only the tiny, dove-like ivory gull and the wedge-tailed Ross's gull, residents of the high arctic that have been sighted elsewhere in Alberta, have not yet been recorded at the lake.

Closely allied to the gulls are the terns, those elegant flying machines that seem to spend their entire lives airborne. Suspended between slender wings, the light body bounces rhythmically as the bird flies over the lake, until it comes to a dead stop, hovering briskly, head pointing down. Suddenly, it crashes into the water and rises immediately, holding a silvery minnow in its scarlet beak. Screaming a nasal triumph, the tern speeds away, carrying its prize to a mate on a distant nest site.

Like fish-eating gulls, the two terns that breed at the lake, the common and the Forster's tern, are white below. They sport a jaunty black cap, but it does not extend below the eyeline. After the nesting season is over, both species retain part of their cap, but the black colour retreats from the forehead, probably in the interest of further enhancing its mimicry.

Interestingly, the third species of tern that breeds in ponds and sloughs all over Alberta, the little black tern, does not hunt fish but feeds on insects year-round. It does not moult into a different winter plumage and evidently it has no need for a white underside.

The largest tern of all, the caspian tern, has not been recorded at the lake in recent times, although Robert Lister mentioned a sighting by Rowan in September of 1923. The caspian tern nests sparingly in the north of the province and might be expected to pass through central regions during migration. A local sighting of this world traveller, that is supposed to look like a medium-sized gull, should be one of the rewards awaiting an alert and lucky gull-watcher in the future.

Chapter 16

OWLS

What makes owls so universally appealing? Is it their round humanoid face in which the eyes are placed forward, like ours, and not to the side as in other birds? To compensate for reduced peripheral vision, owls can turn their heads 270°, observing us over their shoulder in a comical pose. Especially young owls look droll. Like monkeys, owlets make us laugh at a caricature of ourselves.

But perhaps there is more than humor to our fascination with the night-hunters. It is a little startling to discover their hulking shape deep in a tree. The realization that they must have been watching us before we noticed them, gives us a creepy feeling of having been spied upon. Their grave demeanor and nocturnal lifestyle fill us with a sense of mystery and remind us of our innate dread of the dark.

At close range, our largest woodland owl, the formidable great-horned, makes a fierce impression with its glaring eyes and menacing feet that grasp the branch with inch-long claws. For so big and conspicuous a bird, it is surprisingly common at the lake, considering the sparse woods. Every aspen grove, even those without underbrush, seems to shelter a pair. They are year-round residents, but their opportunities for nesting are dependent upon the forced courtesy of other large birds, mainly the red-tailed hawk, that constructs large stick platforms in trees. Owls lack home-building skills and simply usurp a hawk's nest built the previous season. They begin brooding before the red-tails are back from migration, confident of their ability to ward off the returning home-owner if it dares to protest.

The growing owlets soon seem too big for the nest that takes quite a beating, losing one stick after the other, until the structure is weakened to the point where a winter storm may take it down. On the wing before the end of May, young great-horned owls have a long summer ahead to learn independence, although they keep urging their parents to bring them food for a long time. There is no problem if mice, voles, pocket gophers, ground squirrels and snowshoe hares are common. Birds are taken in an opportunistic manner, even during the day. But under cover of darkness, the owls venture into the open to raid a gull colony or terrorize ducks and coots along the lake shore. Leaving only a few feathers at the capture site, the owl carries the carcass back into the woods. No creature smaller than itself is safe from this nocturnal "tiger of the air" that has been known to kill cats and skunks, as well as red-tailed hawks and the smaller owls.

Where the great-horned reigns, other woodland owls are scarce unless the woods offer sufficient cover. At the lake, saw-whet and long-eared owls appear to be very uncommon with only one or two breeding records each. But an open-country species, the short-eared owl, can be numerous. If its staple, the meadow vole, is in good supply and if the winter is mild, some short-ears are present year-round. Partly diurnal, they spring to life from their hidden roost in the grass when the sun nears the horizon. The owls fly about erratically with jerky wingbeats, quick on the upstroke and slow on the downstroke.

Devoid of shyness and caution, they can be lured by imitating a mousy squeak through pursed lips. Hunting by sound as well as sight, the owl will come straight towards us, passing a few metres overhead, round eyes looking down fearlessly. Puzzled or annoyed, it may emit a hoarse cough. Our delight at having these elegant owls so close is tempered with concern for their safety. To try and warn them about the treachery of our kind, we are inclined to shout and clap our hands or to throw stones. But the "dumb" bird does not seem impressed, confident in its wonderful flying powers, out of reach of the clumsy earth-bound creature.

Short-ears display similar impudence in the face of natural enemies. One day, a goshawk emerged from a woodcut, flying low over the pasture to surprise a short-eared owl that had swooped down into the grass. The intended victim rose just in time. It dropped the small prey it had caught and began to climb at a very steep angle, as if ascending a staircase. It stayed well above the hawk that gained height at a slower rate, flying in circles, until it gave up the effort and glided back to the trees.

A deceptively easy target, the slow-flying owls are often attacked by peregrines. One short-ear calmly dodged two falcons that alternately made fifteen violent swoops. But it was like trying to hit a feather with a stick. The buoyant owl, its light body rocking between long glider wings, was untouchable for the powerful enemy. After the falcons left, the owl resumed its diligent quartering of the ground, none the worse for the experience.

Short-ears are excellent hunters. It usually is not long before they make a strike, although they fumble the prey quite often, protected as it is in the long grass. If the hunting is good, owls congregate from far and wide. Without a traditional home range, they are like gypsies, staying where the living is easy. After winters with deep snow that allow the insulated voles to multiply with maximum fecundity, ten or twelve owls can be seen simultaneously over the pastures. The abundance of food brings out romantic urges, expressed during high courtship flights over choice territory. The owls call softly and alluringly. Every few minutes, they descend obliquely for a-ways, clapping their wings below the body which produces an odd ruffling sound. The owls lay clutches of ten or twelve white eggs in simple nests on the ground. Their survival depends on luck, the luck to miss the plow that will come around inevitably as a rude surprise for owls that picked a nest site in the stubble.

If voles remain abundant, a few short-ears stay all winter. In October and early November, they are joined by snowy owls, migrants from the arctic. For the next six months, it is common to spot these exotic white beauties perched on posts along highways and country roads, but they can best be appreciated in the vastness and desolation of the winter scene around the lake. Sitting on a boulder or jumbled pressure ridge off shore, the "ookpik" endures blizzards and intense cold with equanimity, its liquid eyes closed to slits, protected with lashes as delicate as hoarfrost. When it spreads its swan-like wings, its flight is graceful and strong, the picture of nature's artistry in a sterile land that seems devoid of other life. Unfortunately, the thrill of seeing a snowy for the first time diminishes as sightings increase. Their numbers can be impressive if their arrival coincides with an irruption of voles. During spring, as many as 26 have been counted from a vantage point on the northwest corner of the lake. Spotting half a dozen on fence posts or on the ground is commonplace. Through binoculars, in the bright light of late afternoon, the owl's white plumage is visible from a mile away.

If we want to observe the owl's hunting behaviour, we will have to be patient and wait until it flies off. Pouncing into the snow some distance away, the owl bends its head to take a small prey from under its feet. Most often caught are voles and mice, as well as the odd weasel. But after mid March, when ground squirrels emerge from hibernation, the snowy flies low raids over greater distances in attempts to catch the squirrels before they dart back underground. In a similar manner, the

The hardy great-horned owl may be on eggs early in March and the young are on the wing by June. At night, this woodland owl ventures into the open, hunting rodents as well as ducks and coots on the lake shore.

owl may try to take grey partridges that fly up from the ground in the nick of time. Vulnerable to these surprise attacks because of their straight take-off, a partridge may be seized in flight. If the chance presents itself, the snowy also tries its luck at capturing ducks or small passerines.

Instead of these frustrating attacks on birds, why doesn't the owl kill a jack-rabbit, you might ask? The jack or white-tailed prairie hare is quite common on the bleak winter fields, and it would provide the owl with a worthwhile piece of meat. Like all predators, the snowy prefers prey that is easy to kill, and the hare would put up too much of a fight. Its strong hind legs, armed with claws, can do serious damage, and only a desperately hungry owl or an inexperienced immature might risk the struggle. One day, a young female snowy, recognizable by its darkly-barred plumage, had evidently won the battle. It was feeding on the remains of a hare, which it carried along when it flew off, unwilling to

leave it behind where scavengers could find it. Similar ''tight-fisted'' behaviour is seen of great-horned owls, which will carry half-consumed prey to their winter roost, saving the food for the next meal. It is a habit that owls have acquired to better cope with the lean larder of their northern environment.

During April, when snowy owls stop over at the lake on their way back to arctic tundras, central Alberta is usually snow-free, rendering the white birds very conspicuous. Curiously, they seem to have an affinity for the last drifts or any other white object. Seven owls were resting on or next to a white drainage pipe. Another sat between the bleached bones of a cow. But most will spend the day on the still frozen lake, snoozing on the ice far from shore. Towards evening, they fly to the land or rise into the bright northern sky, resuming their voyage to the arctic and leaving the lake shore to their cousin, the great-horned owl, whose hooting can be heard from a woodlot a little inland.

The earliest fall arrival date for snowy owls is 2 October 1974. The earliest record for rough-legged hawks fell in the same year. Perhaps, these two arctic predators were forced to leave their tundra summer range earlier than usual because of a scarcity of lemming prey.

Chapter 17

SONGBIRDS OF OPEN COUNTRY

The passerines, informally called songbirds or perching birds, make up by far the largest avian order, containing 21 families with a total of about 230 species in Canada, almost half of the 500-plus birds that have been identified within this country's borders. At the lake, we can list about 126 passerines that either nest locally or travel through on their way to more northerly breeding grounds.

The local regulars can be split into two groups according to their general habitat preference, either open or treed, although there is some overlap and seasonal variation. Typical songbirds of the open are larks, pipits, buntings, longspurs and blackbirds. Typical woodland families include thrushes, warblers, flycatchers and vireos. Families with one or more representatives in both habitats are corvids, sparrows and wrens. Nearly all of these passerines are highly migratory and occur at the lake only from spring to fall, arriving and departing unseen, travelling at night.

As mentioned before, the first migratory open-country bird to return to central Alberta is not a hardy goose or hawk, but a tiny passerine, the horned lark. It is back even before winter is officially over, flitting along the country road or from one snow-free patch to another in late February or early March. On mild days, its tinkling calls complement the rushing sounds of meltwater.

Unfortunately, it appears that these pretty larks have become quite scarce around the lake. During a ten-kilometre walk along the west shore, the watcher rarely flushes more than one or two. Perhaps this local scarcity is tied to subtle changes in habitat and land-use that do not suit the lark's needs. It prefers short-grass terrain. Fact is, that the local pastures are less heavily grazed than formerly, leaving the grass quite long, especially after a wet growing season. Another short-grass passerine, the chestnut-collared longspur, used to breed in the 1960s but has rarely been seen since around the lake.

Somewhat later in spring, in April and early May, horned larks suddenly become more numerous, roaming about in flocks. Smaller and darker than the pale local subspecies, they are northern migrants on their way to the tundra, often in the company of other arctic travellers such as Lapland longspurs and snow buntings. Restless and often disturbed by hawks, longspurs may fill the sky in their thousands, even tens of thousands, streaming overhead and twittering a communal flight song of great charm. The flocking of snow buntings is spectacular to watch as the black-and-white birds gyrate in tight formations to escape from a falcon that scares them up from the fields. If they have been harassed several times, the buntings become increasingly irritable, often flushing for no apparent reason. When they descend to resume feeding, the flocks roll over in a peculiar manner with the last birds passing over and landing beyond the vanguard.

On the ground, obscured in furrows or stubble, buntings and longspurs are difficult to see, but perched on a fence wire, side by side in long rows, we can observe them at leisure through binoculars. With luck, we may find a

Smith's longspur, the prettiest arctic passerine and by far the most talented songster. Once we become familiar with its call note, less harsh than the Lapland longspur's call, we will see it more often, identifying it readily when it flushes from the ground. In flight, the two white wingbars and the buffy undersides are characteristic of this much sought-after species which is more common than is generally thought. Smith's longspurs appear to stay together in pure flocks that do not mingle with other longspurs. The flight call, wingbars and buffy undersides remain helpful features during fall when the Smith's longspur is in winter plumage and much rarer at the lake than in spring.

While the arctic passerines are moving through, locally nesting songbirds are calling attention to themselves with territorial displays. The mellow phrases of the western meadowlark are heard all over the pastures, and where the grass is rank, the brightly coloured bobolink flutters into the sky, bursting out in a delightful tune.

Flitting furtively over the ground or between bushes, a variety of sparrows challenges the best of birders. The savannah sparrow likes to perch on a low branch or fence post before giving its lisping song. A quick glance through the binoculars is sufficient to recognize this common species by the yellow stripe above the eye. But other grassland sparrows are much harder to identify and require persistent effort. In late May and June, the sounds of LeConte's and sharp-tailed sparrows are a very soft, insect-like buzzing that some people are unable to hear at all. Walking slowly into the direction of the sound, through rank sedges over wet ground, we try to flush the little birds, hoping that they will pause in a bush. It takes skill and patience to spot the sparrow between the leaves and to focus the glasses quickly enough for identification. Both the LeConte's and the sharp-tailed are exquisitely coloured with yellow and buffy tones on face and undersides, but diagnostic differences between the two require careful checking with the field guides.

Notwithstanding the fact that most birdwatchers seldom if ever obtain a good look at either one, these sparrows are far from rare at the lake. In 1987, in a survey of 11 hectares of marsh grass studded with low willows, just west of the weir, Roger Jones located a total of 8 singing male LeConte's and 6 sharp-tails, as well as 15 savannah sparrows.

Two other sparrows are regulars on the edge of the grasslands, along the ditches and in the trees, the clay-coloured and the vesper sparrow. We are alerted to the presence of the first by its "song", a series of low buzzes. But the vesper sparrow produces a sweet and mellow phrase ending in a jumble of twitters.

Birdwatchers familiar with the sounds of southern prairie passerines occasionally hear the Baird's sparrow that frequents short-grass pasture. Its song sounds like three or four "zips" followed by a musical trill. To obtain a good look at one of these nondescript brownish birds, we must persist and wait for a chance to catch it in the open.

Another short-grass passerine with a spotty distribution around the lake is the Sprague's pipit, and it too can best be located by listening for its characteristic song. Given from high in the sky, it sounds like a thin metallic sawing on a descending scale: "ching-a-ching-ching." Scanning the sky overhead through binoculars, we discern the pipit as a tiny dot, fluttering in circles over its nesting territory. At intervals, it folds its broad wings to glide a little ways, at the same time producing its jingle. Eventually, its ardour abated, the bird drops like a stone to the ground, no doubt meant as a show of bravado for the benefit of its brooding mate.

Locating the nest of a pipit or any other grassland passerine is a matter of careful search or sheer luck. Sooner or later, the observant hiker will notice a lark, pipit or sparrow slipping away almost underfoot. Deep in a tussock, we discover a cosy basket of woven plant material. Of course, it is best to go quickly on our way. But who can resist a brief look at the eggs or tiny young? Surprisingly often there is evidence that a cowbird found

On the way to its nest, hidden in the grass, a western meadowlark seems to pause for the photographer.

Nest parasitism by cow birds appears to be very common at the lake. At right is a nest of a LeConte's sparrow containing three larger cow bird eggs. Above, a savannah sparrow (note the band on its leg) removes a fecal sack excreted by one of its monster foster chicks.

the nest before we did; one of the eggs is noticeably bigger than the others. This cowbird egg will hatch into a hungry, demanding chick that will take the rightful place of the other young birds.

Nest-parasitism, similar to the notorious practice of the European cuckoo, appears to have arisen in the brown-headed cowbird out of its need to travel and keep up with the herds of bison roaming the western plains. In this day and age, when the shaggy beasts have been replaced by domestic cattle, the cowbird seeks the company of Herefords and Holsteins that attract and disturb insects on which the cowbird preys. Although it may spend all spring in a single pasture, eliminating the need for a nomadic lifestyle, the bird persists in forcing its monster progeny on tiny foster parents.

Cowbirds belong to the blackbird subfamily, which is unique to America and well-represented at the lake with colourful and melodious performers such as meadowlarks, orioles and bobolinks. But the group is named after six members that are mainly black and far from a delight to the ear when they squeak, gurgle and stutter their spring cacophony. Brassy and glossy, the common grackle is the largest of the clan, flaunting its wedge-shaped tail in a woodlot here and there, often near farms. The brewer's blackbird looks sophisticated with its shiny plumage and silver eyes. It nests in small colonies in buckbrush on the pastures. But by far the most prominent family members are the red-winged and yellow-headed blackbirds that decorate the reed beds. The yellow-headed takes up residence in the heart of the marsh where the cattails are tallest and the water deep, while the much more numerous and adaptable redwing takes anything it can get, including some dryland habitat.

Sometime in early April, the adult males arrive in the winter-ravaged reeds, and immediately proclaim property rights with frequent "song". The yellow-headed utters a weird discordant croak, but the redwing's shrill "onk-a-reeee" has a heart-warming quality that we have come to associate with springtime in the marshes. Of course, to the birds themselves, their calls have nothing to do with musicality or entertainment. They are meant to challenge and intimidate rivals intent on crowding out the cock-of-the-castle. As an emblem of authority, the redwing flashes his shoulder badge, opening the crimson patch as wide as possible by spreading his wings. Without this badge, the redwing would have neither status nor property rights, and hence no chance to obtain a female. Juvenile and yearling redwings that do not have distinct red epaulets are forced to remain bachelors for one or two years.

About two weeks after the males have established themselves, the ladies arrive. Much different from the male, the female redwing is small and brown with white and dark streaks. Wherever she lands in the reeds, she is immediately claimed by the male of the territory, who also welcomes any other female that happens to land in his domain. The females fight among themselves, but their territorial needs are smaller than the male's, and the bigger the boss's holdings, the more households he can accommodate and the more offspring will inherit his genetic qualities. According to researchers such as Robert Nero, some male red-winged blackbirds collect in excess of ten females, but most have less than four.

Given the competition and constant bickering between the sexes, the colony seems in a perennial state of uproar, but eventually the females get down to weaving soft baskets suspended between cattail stalks. Ten to twelve days after brooding begins, the eggs hatch and the young are fed a high-protein diet of insects by both parents. Of course, the male will have to divide his attentions between several households and the more energy he invested in acquiring a large estate, the more he will be "taxed".

In July, after the young have fledged, the families abandon the reed beds and travel to the fields, where they change their menu to weed seeds as well as grains. The latter may incur the wrath of farmers, particularly if bird

99

numbers grow to thousands. Each evening, the scattered flocks converge from far and near to traditional sites along the lake where they spend the night in reed beds, settling down after much noisy careering and manoeuvring. Such communal roosting sites in the United States have contained over one million birds. At a total estimate of 400 million, the red-winged blackbird is considered the continent's most numerous bird.

The nightly gatherings are believed to serve several purposes. Firstly, there is safety in numbers. The individual's risk of being killed by a hawk or owl becomes less as the number of blackbirds in the flock grows. Secondly, as recent research on European starlings has divulged, the evening rendezvous provides a chance to communicate foraging opportunities. Birds that are still hungry at the end of the day recognize and follow well-fed individuals to better feeding grounds next morning.

Foraging flocks of mixed blackbirds often include starlings and associate loosely with crows, which have been roaming the countryside since August. During October, the black crowd vacates central Alberta, leaving just the resourceful magpie as the only open-country ''songbird'' that is content to remain with us all winter.

It takes skill and patience to obtain a good look at two confusing open country sparrows: the LeConte's (top) can be told from the sharp-tailed (bottom) by the narrow white stripe on the crown and the dark streaks on the side of the body.

Although it breeds commonly in lake-side shrubbery, the yellow warbler allows the casual birder seldom more than a fleeting glimpse.

Chapter 18

WOODLAND SONGBIRDS

The aspen and willow brush flanking the lake's south and east shore may not seem the richest habitat for songbirds, but there are a few surprises. After the termination of the grazing lease in the Natural Area by Lister Lake, the woods have developed a rank understory that offers a much improved habitat for birds as well as mammals, despite the sandy soils that underlie the region. In July of 1987, the fluted song of hermit thrushes was common in the dank brush east of the weir, alerting bird-watchers to the presence of a species that is generally considered synonymous with remote wilderness habitat.

Other songbirds, in the truest meaning of the term, that nest in the woods by the lake include the ubiquitous American robin and the odd veery. Rivalling or even surpassing these thrushes, depending on one's point of view, is the warbling vireo, whose exuberant performance is certainly superior to the monotonous phrases of its close cousin, the red-eyed vireo.

The highest stands of poplar resound with the whistle of the northern oriole, its bright voice a match for its gorgeous orange and black plumage. Also colourful but more subdued in voice is the yellow warbler, which occurs throughout the woods, including the low willows by the marsh where the yellowthroat gives away its hide-out with emphatic calls of "witchery, witchery."

The above two warblers are the only common ones known to nest at the lake, but a surprising variety of migrants have been banded, including the black-throated green warbler. Since its establishment in 1983, the Beaverhill Bird Observatory (BBO) has caught a total of twenty-three species of warblers in mist-nets set up between the bushes near the shore. The grand total of all species of birds captured by the BBO, as of 1997, is 101.

Most often banded were least flycatcher (4008), yellow warbler (3319), yellow-rumped warbler (2373), Tennessee warbler (1258), and black-capped chickadee (724). The latter is the only songbird that stays in the area all year, delighting us already in February with its whistled territorial calls. Formerly, chickadees found little opportunity for nesting in the young woods where cavities of sufficient size are scarce, but the little birds have taken advantage of nest boxes placed by BBO people to attract bluebirds, tree swallows or house wrens.

Common farther inland, bluebirds seldom occur close to the lake, but swallows have multiplied in a big way. They readily adopted about 150 nest boxes placed on fence posts along country roads, in the marsh and along the lake shore by the BBO station, where ornithologists Geoffrey Holroyd and Peter Dunn set up an experimental study of swallow productivity and mating behaviour. In particular, they wanted to investigate the role of food supplies in the incidence of bigamy. In a Ph.D. project for the University of Alberta, Dunn hypothesized that the abundance of insects on the lake shore would allow male swallows to defend more than one nest box and adopt more than one brooding female, in contrast to the swallows along the road where food supplies were believed to be more limited. To measure the relative abundance of flying insects, Dunn

103

placed pole nets, shaped like a wind-sock, in the two areas. Although lake-side nets collected over 3,000 insects on peak days, there were also times when they contained very little. Similarly erratic food supplies were measured along the roads, making a clear comparison impossible or invalid. Nevertheless, lake-side swallows produced more eggs per nest than those along the roads, and seven of the males were bigamous and attended two females each.

Male bigamy was also found among house wrens studied by Mike Quinn in a Master's Degree project for the University of Alberta. He placed 208 nest boxes in the woods east and west of the weir. The majority were occupied by wrens, and at least eight of the males were polygamous. Competition and strife between their females were avoided by timing. At the start of incubation by the first female, the male would expand his territory to include a second box some distance away. By the time his second mate had laid her six to eight eggs and began brooding, the first clutch was hatching, and the male returned to assist female number one with

feeding her brood. He switched his attentions again when the second brood hatched some 14 days later.

During his three-year study, Quinn banded 750 house wrens, including 200 adults. Not one banded young returned the following season, and only four of the adults came back to the study area next year. Incidentally, none of the banded wrens was reported recovered from outside the area. Of all passerines banded by BBO, up to 1997, only five have been recovered outside Alberta.

An unexpected bonus of the various BBO surveys was the discovery of several rare birds nesting in the area. Besides the singing hermit thrushes, that may have been nesting, BBO people happened to locate a nest of a veery, a mourning warbler, and two nest each of black-billed cuckoos and sharp-shinned hawks.

As the aspen woods in the Natural Area continue to mature and develop, undisturbed by cattle, we may expect more ornithological surprises in the future.

During spring and fall, at least 23 species of warblers have been netted and banded at the lake, among them the brilliant magnolia (top) and the Canada warbler (bottom).

Chapter 19

RARITIES

In the vernacular of birdwatchers, rare birds are species considered absent or scarce in a particular locality or during a certain time of year. The degree of rarity varies widely and is subject to review. A bird may be rare on a seasonal or regional basis only. For instance, in central Alberta, robins are rare during winter but abundant from April to September. During summer, lark buntings are rare at Beaverhills Lake but common in the south of the province. Other species that have been classified as local rarities, are rare across Canada or in all of North America, such as the Eurasian ruff and the Siberian sharp-tailed sandpiper. Part of the attraction of birdwatching is that surprises are possible in all categories of rarity. As long as birds have the awesome ability to cross oceans and mountain ranges, they are liable to show up anywhere, anytime, as they have proven time and again to the delight of birdwatchers across the globe.

In the field, certain bird species are more elusive for some people than for others. Is the peregrine falcon rarer than the sharp-tailed sparrow? The majority of birders would probably affirm; peregrines are not often reported, whereas the sparrow is a regular that can be found if you know where to look. Yet, this author has never enjoyed a no-doubt-about-it look at a sharp-tail, but he has observed hundreds of peregrines. What we see depends on where we set our sights, on the horizon or on the grass at our feet.

Rare sightings have a lot to do with the watcher's expertise and experience, or the lack of them. Unaware of the common occurrence of white pelicans, novice birders have reported the improbable sight of whooping cranes soaring over the lake on their great white wings with black tips. Others have mistaken common Bonaparte's gulls for mew gulls because their plaintive, mewing calls seemed to give them away. Misidentification happens even after the most careful scrutiny. One neophyte ornithologist had his telescope trained on an immature Bonaparte's gull swimming in a pool nearby and declared it, after much checking in his spotless field guide, a red phalarope in winter plumage!

Of course, beginners are entitled to make mistakes, but their ''rarities'' should not be reported in newspaper bird columns or club newletters unless they are checked out by an expert, who has made his own mistakes in the past and is now aware of the potential for confusion.

To reliably identify a rarity, it helps if the watcher is familiar with the species from elsewhere, or if one has been alerted to its unusual local occurrence by other observers. In many cities, birdwatchers organize so-called rare-bird-alerts or hotlines, which can be contacted by telephone for up-to-date information. If other birders have a chance to confirm the correctness of the original report, the identification becomes more acceptable to those who compile checklists and write handbooks. As any expert will agree, scepticism about reported rarities is a must, but at the same time, one should have an open mind about the unexpected. After all,

Photo: Dick Dekker

This long-tailed jaeger was seen by several parties of birdwatchers between 8 and 12 September 1977. It captured and ate at least four lesser yellowlegs. Quite unafraid of people, it allowed the photographer to approach closely.

beginner's luck *is* possible, and it can be frustrating for a talented novice to report an unusual bird only to be met with disbelief or ridicule.

In the final analysis, birdwatching should have its own rewards, regardless of what others think we saw. Some of the most ardent amateur ornithologists have chosen to be loners, derisive of the social rewards of group excursions, and ignoring rare-bird-alerts and hot tips. No doubt, discovering one's own rarities offers the greatest thrill. But communicating one's findings, even well after the event, helps other birdwatchers become aware of the possibilities. Yellow rails and sharp-tailed sandpipers might still be unknown at the lake, were it not for someone's willingness to share information at meetings and in the literature. The keenest birders specialize on certain favourite groups, such as waders, passerines, or raptors. Few excel in all areas, but together these specialists complement each other, for mutual benefit and enjoyment.

As the number of qualified observers grows, the list of rarities seen at the lake becomes longer with diminishing chances for adding more new species. However, surprises keep turning up at the rate of one per year or so. Between 1975 and 1988, at least a dozen "firsts" were reported, among them little gull, kittiwake and red-throated loon.

By far the majority of rare birds at the lake are pelagic or arctic species that normally occur along the coasts and seldom stray inland, except when storms push them off-course during their migration. Pacific species that have been recorded only once, thusfar, are surfbird, wandering tattler and ancient murrelet, while the western sandpiper has shown up a dozen times. A fascinating pelagic brigand such as the jaeger was proven to be much less scarce than previously believed. Since 1964, the author has recorded 141 jaegers on 61 dates. Although some of these birds stayed too far off-shore to see details of their plumage, others were observed at close range and could be positively identified by

their elongated central tail feathers as parasitic jaegers. Light-phase adults were most common but three dark-phase birds were seen, as well as half a dozen mottled immatures. Often two or three were flying together, and once, five jaegers were in view simultaneously. Although they showed up in any month from May to October, 99 out of 141 did so in September.

The best way to discover a jaeger is to keep an eye on the gulls. If resting ring-bills or Bonaparte's suddenly rise, look beyond them for the tell-tale silhouette of the jaeger with its characteristic dark wings and white belly, diagnostic from a kilometre away, long before you can discern the white flashes in the wings or the protruding tail feathers. Most likely, the jaeger will have singled out one of the gulls that screams in terror and tries to dodge the jaeger's rushes by veering and twisting. The pursuit may last a few seconds or several minutes, with the jaeger intent on forcing the gull to cough up the contents of its crop. If pressed hard, the gull will regurgitate promptly. Others do so only after persistent prodding. The pirate dives after the falling food and catches some of it in mid-air.

Less commonly witnessed is the jaeger's expertise at hunting small passerines and shorebirds. A pair may cooperate deftly in chasing down a phalarope or sandpiper until one of the super-charged hunters catches the prey in its bill and carries it down to the water. Few of their chases fail, unless their intended victim plunges into reeds. The only long-tailed jaeger seen at the lake seized at least three lesser yellowlegs and pecked them to death while standing on the prey in shallow water.

Of late years, for some unknown reason, jaegers have become less regular at the lake. From a high of 0.41 sightings per fall day during the five-year period 1964-1968, mean sightings have dropped to 0.16 and 0.08 respectively in 1974-1978 and 1979-1983. Since 1983, the author saw jaegers only in 1985, although others reported one or two in 1987.

Besides the many rarities that originate from Pacific and arctic coasts, the lake occasionally

Two pairs of black-necked stilts nested at the lake in June 1977, the first successful breeding of this southern species in Canada. As noted by the photographer, the birds always faced into the wind.

receives an out-of-range visitor from southern or central regions. In the spring of 1976, a great egret paid a conspicuous but brief visit to the lake's northwest corner, and in 1987 one of these white herons showed up in the Amisk Creek marshes where it was photographed by landowner Lawrence Kallal. A smaller species of white heron, either a snowy egret or a cattle egret, was seen flying along the north shore on 19 May, 1974. It was later noticed by farmer Marshal Eleniak, who saw it standing on his pasture while he rode by.

A real southern beauty, a male wood duck, was recorded in May of 1985, probably blown north on a strong wind that forced the migrating bird to travel farther than usual. Long-billed curlews, a species of southern prairies, quite often overshoot their home range during May, as they have been seen at the lake more than a dozen times. Although one of them stayed around until June 18, all others disappeared soon, probably backtracking to more traditional range.

A truly spectacular southern dandy that came to stay for the season was the black-necked stilt, which is perhaps the lake's most celebrated rarity that did us the added honour of raising its young here, establishing the only confirmed *successful* nesting of this species in Canada. (There are two nesting records for Saskatchewan, but the eggs did not hatch.)

The first stilts were reported at the lake on May 1, 1977, when a pair was seen feeding among avocets in shallows near the dam. Three weeks later, the author observed two stilts at the north end, and on June 4, he found four of them on an inundated field along the west shore. When he returned the next day, accompanied by Robert Lister and ornithologist Dr. Chip Weseloh, four stilts were still at the same site. At the approach of the party, one of the birds ran from behind a clump of marsh ragwort in an agitated manner and resorted to distraction display, dragging its wings. Twenty minutes later, the intrepid Weseloh, slogging on bare feet through mud, had found Canada's first nest of the black-necked stilt! It contained four eggs, laid

on a shallow mound of debris and plant material surrounded by water. On a return visit on June 17, Weseloh located a second nest, a little distance from the first, containing seven eggs. Subsequently, during later checks of the nests, at least four of the eleven eggs had hatched and two young were banded. Two other eggs probably hatched, but three were addled and collected to be preserved in the Alberta Provincial Museum. Two other eggs had been taken from the nest earlier.

When Robert Lister wrote about the stilts in his bird column, then a weekly feature of the Edmonton Journal, his report created quite a stir in birdwatching circles and many people wanted to see the birds. However, to minimize the risks of further disturbance, the party of discoverers had agreed to keep the location a secret. Eventually, half a dozen friends were taken to the nest, while many requests were refused, which caused some resentment that has not yet been forgotten in some instances. Ludo Bogaert and Peter DeMulder had also been denied detailed information but learned that the stilts were on a field along the west shore. They decided to search for themselves and were successful in finding the nests. Subsequently, Ludo made the excellent pictures that accompany this chapter.

Yet, the legitimate question remains: should all necessary disturbance have been avoided from the start? Does the discovery of rare birds, especially of species that might be on the verge of expanding their nesting range, carry with it the responsibility to keep the location a secret, at least until the risks to eggs and small young have passed? The protection of nesting sites of rare birds continues to create controversy in western Europe, where bird photography has increased in popularity by leaps and bounds. To discourage the photography of nesting birds, some European nature magazines refuse or limit the publication of such material.

In the case of the stilts, there is no question that the nesting was a success despite the frequent visits by people. Ironically, if these birds had escaped detection, the nests would not have fared so well. They would have been destroyed unwittingly by the farmer, who began plowing his field as the soil dried out. Just before he was to start on the section that contained the nests, Weseloh and the author alerted the farmer and showed him the rare nests. On June 24, Bruce Stadnik cultivated his entire field with the exception of a small area, the size of a table top, around each nest. And the birds took the disturbance all in their stilted stride!

Soon after the young hatched, the adults led them into an adjacent marsh where they were last seen on July 9. The obvious question on everybody's mind was: would the birds return next spring? However, only one sighting was reported in May of 1978, but no nests were found. No other stilt records exist for the lake up to the time of this writing.

The reason for the temporary appearance of the stilts is open to speculation. As has been noted before, the early spring of 1977 was marked by wide-spread drought conditions in the northwestern United States. The disappearance of traditional nesting sloughs in these southern localities may have forced the stilts north to Alberta. They found ideal conditions at the lake since water levels were dropping, creating optimum mudflat habitat and rich food supplies. In 1977, the number of avocets was also exceptional with 120 in view simultaneously at a point on the west shore.

After drought conditions eased in its traditional range, the stilt found no more need to travel north, surrendering its brief status as a famous lake-side resident, but retaining the uncertain classification of a local rarity, albeit one of special distinction.

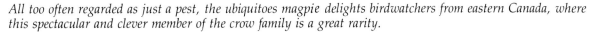

All too often regarded as just a pest, the ubiquitoes magpie delights birdwatchers from eastern Canada, where this spectacular and clever member of the crow family is a great rarity.

Red foxes are locally quite common along roads and near towns, but they avoid the pastures around the lake, probably because of competition from the much larger and aggressive coyote.

Chapter 20

FROM BEARS TO WEASELS

Except for gopher shooters or coyote hunters, few people go to Beaverhills Lake for the specific purpose of looking for mammals. Yet, the list of quadrupeds that occur around its shores is neither short nor lacking in variety, and may contain some surprises.

In June of 1987, two visiting birdwatchers from Texas saw a cougar chasing a deer among the willows near Lister Lake! The lack of details of this exciting report leaves room for scepticism, although the occurrence of cougars in the area is not impossible. After all, the big tawny cat has been seen occasionally in the Beaver Hills and there are even reliable reports from the outskirts of Edmonton. Apparently, cougars roam widely and stalk their prey wherever they find it. And there is no lack of deer around the lake.

Other unsuspected but unsubstantiated local sightings feature wolverine and otter, species that are far less likely to show up in central Alberta than cougars. And one day, years ago, on the muddy margin of Lister Lake, the author found the five-toed, five-centimetre wide tracks of a mustelid he would have identified as a fisher, had the location been in the Edson forest where fisher tracks can be expected. At the lake, one needs to have more solid proof of its occurrence than a few likely tracks. Interestingly, in March of 1990, the Alberta government released 17 fishers in the Beaver Hills just west of the lake.

Bears seem improbable too in the lake's agricultural surroundings. Yet, in the summer of 1972, a local farm lady experienced a real scare when a black animal rooting around in her vegetable garden was not a calf, as she thought when she left the house to shoo the marauder away, but a real bear! Subsequently, a black bear was spotted by an airborne wildlife officer somewhere east of the lake. Black bears have occasionally been sighted in the Beaver Hills, and at least three were shot near Edmonton in the last ten years.

It is the author's silent hope of meeting a wolf one day, while hiking along the secluded east shore. The chances should be slim, as there are very few reports, let alone reliable ones, of wolves in the Beaver Hills. However, as long as mammals have four legs, they are capable of showing up in unexpected places. Dispersal of yearling animals is especially prevalent in wolves. Tagged individuals have been recorded as far as 800 km away from their place of origin. And Beaverhills Lake is no more than 200 km from traditional wolf country to the west and north.

The largest wild canid common to the Beaverhills region is the coyote. It is frequently seen since it is often abroad during the day, skulking along the lake shore, perhaps doing a bit of "birdwatching" in its own way. In turn, some birds watch the coyote, with curiosity and apprehension. Ducks often follow it or swim towards it, a curious compulsion exploited formerly by Dutch market hunters, who used a small yellowish dog to lure ducks from a pond into a reed-enclosed funnel leading into a wire-covered enclosure. The age-old design of this "eendenkooi" was copied for the famous Delta waterfowl research station in Manitoba, where ducks are captured for banding purposes.

Coyotes and other wild predators capitalize on the ducks' curiosity for their own benefit. Feigning disinterest or sleep, the coyote waits for the prey to come close enough for a flying leap. One fall day, a coyote was observed stuffing a snow goose under a clump of reeds, perhaps caching it for later use. Whether the goose had been acquired dead or alive was not known. Of course, coyotes stalk and hunt geese whenever the opportunity presents itself, and the risk of canid ambush has made the goose an exceptionally watchful creature with a strong preference for open surroundings. Even swans may occasionally be captured. Tracks indicated that a coyote had raced out of willows into shallow shoreline water to seize a whistling swan, one of thousands that each October collect on the lake and feed on the roots of aquatic plants. A trail of white feathers led across the muddy shore into the bushes. Unless the swan was a cripple, its slowness in taking off had probably predisposed the great, ponderous bird to the swift attack that may have taken place during the night, while the swans were asleep.

The coyote's year-round staple prey is not waterfowl but rodents. It catches meadow voles with ease, pinning them down under its front feet after a precise but comical-looking pounce. In times of plenty, the coyote plays with its food, kicking the captured vole across the snow-drifts like a soccer ball, or flinging it up repeatedly into the air.

When small prey is in short supply, packs of coyotes become persistent predators of deer, especially those that are weakened by severe weather conditions. However, the coyote's mainstay during winter are the carcasses of cattle dumped in the bushes by farmers. Scavengers congregate from miles around and as many as ten or twelve coyotes have been seen in a single field. Such sightings used to spark calls to the government predator control officer, who would set out poison baits. But after the proliferation of the snowmobile in the late 1960s, coyote populations seem to be hunted down to lower levels than before, making control by poison quite unnecessary. Low densities tend to benefit a species that is prone to terrible contagious diseases such as sarcoptic mange, an affliction of the skin that results in partial or complete loss of hair. Prior to the 1970s, before snowmobiles became in common use by coyote hunters, it was routine to see animals with ratty, hairless tails and nearly bald bodies. The disease seems less prevalent today.

Another serious menace to the coyote was the revival of an ancient hunting method, the coursing with hounds, which became suddenly popular in central Alberta during the 1970s. The trained killer dogs were trucked around the country roads and let go in pairs when a coyote was sighted. The hunters followed the chase by road and released more hounds from the truck as the occasion demanded. Some of these men boasted of killing more than a hundred coyotes per winter, during a time of rising fur prices. Eventually, the Alberta government outlawed both the practice of coursing and the running-down of wild animals with snowmobiles. However, the proliferation of all-terrain vehicles, including balloon-tired trikes and bikes, continues to give the modern coyote-getter an unprecedented mobility and advantage over its quarry. Appalled by the slaughter, some local landowners have become quite protective of their coyotes, which are valued as allies in the battle against mice and ground squirrels.

An interesting phenomenon that coincided with the downtrend in coyote populations was the expansion of red foxes, which were virtually absent from central Alberta prior to the 1970s. It is possible that the local demise of its larger and aggressive cousin allowed the fox to establish itself in the vacant canid territories. There is no doubt that the coyote is the arch enemy of the fox, and it appears that the two competitors are segregated spatially. Foxes are restricted to a few locations near farms and roads, where they seem to seek the involuntary protection of man. In 1987, a pair of foxes raised pups in the town of Tofield on

A few badgers manage to hang on around the lake despite active opposition from landowners who object to the animal's compulsive digging.

the edge of the hospital parking lot. The author has yet to see a fox on the coyote-dominated pastures around the lake. In twenty years of checking for tracks along muddy shores, he has found fox prints only once, on a reedy section of the east shore. Fox tracks can be identified positively by the small toe pads and the transverse ridge of callous across the heel. (For more details on the interaction of foxes and coyotes, the reader is referred to *Wild Hunters*. Please see literature list.)

The coursing of hounds, when it was still legally practiced in the region, not only resulted in the destruction of numerous foxes and coyotes, but also in the local disappearance of the lynx. Unobtrusively, as is the cat's way, a few lynxes had established themselves along Amisk Creek, where farmers had left some poplar groves around wet spots in the grain fields. Untouched by cattle or fire, the rich understory provided habitat for varying hares and ruffed grouse. During winter, the big pads of the lynx could often be found along Amisk Creek and by Lister Lake, and occasionally anywhere around the main lake. The hounds treed a few lynxes, others were shot near farm yards. Fact is, that the species became practically extinct in the area by the late 1970s. The hounds have been called off for quite a few years now, and the lynx may return. However, most of the woodlots that used to contain pockets of hares have since been bulldozed and burned to make room for more agricultural production.

One dramatic meeting with a lynx near the lake is preserved as a treasured memory. Somewhere in the willows, coyotes were yapping in the excited way that meant alarm, usually indicating the presence of larger predators such as wolves or bears in wilderness environments. What could have upset the coyote on these farmlands? A little later, a lynx emerged from the bushes. Ignoring the coyotes, it strode along calmly on a course that eventually brought it quite close. Just before it re-entered the woods, the author sounded a mousy squeak. The lynx halted for

a moment and gazed at the source of the sound in an unscrutable way, without fear or concern. It then plodded on through the deep snow. Its large, furry feet looked like dustmops.

Compared to the felid and canid families that have only one or two members each at the lake, the mustelids are well represented with six carnivores of varied looks and habit. In descending order of size they are: badger, skunk, mink, long-tailed weasel, ermine and least weasel.

Sand-coloured and squat, the badger may move about in the grass unnoticed, but its presence is betrayed by the oval-shaped burrows in the pastures. When you walk quietly along the fence lines on higher ground away from the shore, especially in early spring, you may surprise a badger at work, while he kicks up sand and earth in forceful bursts, half a metre high. Presently, he will stop and look about warily before resuming his digging. Sometimes the badger is so engrossed in excavating its prey that it is possible to approach closely. When it finally emerges from the burrow, it may hold a pocket gopher or ground squirrel in its mouth. If it becomes aware of you, it will growl explosively and vanish underground in a hurry, plugging the burrow behind itself with loose soil, as is the animal's habit before retiring.

Badgers would be quite common around the lake were it not for the active opposition from farmers, who object to the axel-breaking holes in their fields or pasture. The animals are shot, trapped, or run over with tractors, but they manage to hang on precariously, unless people destroy and poison the badger's staple, the ground squirrel.

Farmers are equally opposed to skunks on their property, but for entirely different reasons. The black and white scoundrels have a habit of denning under barns and granaries, risking nasty surprises for humans. The sight of a skunk at close range, spray gun at the ready, never fails to incite panic. Moreover, there is a chance of encountering a rabid animal, since skunks are known to be carriers of the dreaded disease.

Of course, skunks are interesting in their own right, and despite man's hatred and persecution, they are not at all scarce around the lake. Like the badger, the skunk hibernates during winter, but it emerges early. As soon as the weather warms in March, you can find its small bear-like tracks in snow or mud on the fields as well as along the shore. Skunks do not mind getting wet and they travel commonly between the reed beds in search of dead and living food items, including decaying carcasses of birds and mammals. One ate from the remains of a porcupine. Another was feeding on dead lady bugs washed up on shore, covering the ground like a red carpet.

Skunks spend a lot of time rooting out beetle larvae and grubs from under the grass. They can be deeply involved in their digging and pay little attention to the surroundings, so that you can walk up quietly. All the while, the animal holds its conspicuous tail high over its back, for all to see. Why does it advertise its presence? Does the tail act as a warning flag? Or as a decoy that would draw the attack from a predator? Also foxes hold their tail straight up in the air when they are preoccupied with small prey in the grass. A fox's tail, with its white tip, is similarly conspicuous as the skunk's flag. A predator such as an eagle might mistake the tail for an animal with its head held up in an alert pose that should discourage attack since predators prefer to attack from behind and by surprise. If the eagle does make a bold move, it would only seize the prey's least vulnerable part. No doubt, any animal that tackles the tail of a skunk has a surprise coming!

If confronted, skunks stand their ground bravely and may even take a short run towards a person, with startling effect! However, few people have actually been sprayed. Even if teased, skunks tend to hold their fire, although a faint whiff of the musky smell might hang over the area afterwards, not obnoxious at all, perhaps even rather pleasant and typical of fall odours around the lake.

The skunk's reputation as a nest raider brings it into conflict with hunters, who usually shoot the animal on sight. There is no doubt that the skunk likes the occasional egg, but how often does it indulge? It is rare to catch it in the act. One skunk was observed to consume four pintail eggs out of a clutch of ten, but no check was made to see whether it returned to destroy the remaining eggs later, or whether the duck resumed brooding.

The next member down the line of the weasel clan is the mink, which is of regular but sparse distribution at the lake. Its tracks, in the familiar two-by-two lope pattern, may be found during fall and early winter, but the animal itself is rarely seen. On a day when the lake was solidly frozen, a mink climbed a tree and looked down boldly at approaching people. Another mink had obviously not been so alert in escaping from danger; freshly killed, its body was lying in the snowy woods, surrounded by the tracks of coyotes and a lynx.

The shallow lake, that probably freezes to the bottom each winter, is perhaps not a suitable habitat for the mink, which is primarily an animal of wooded stream banks, where burrows and natural openings allow access to the running water beneath the ice. During the course of winter, dropping creek levels create breathing and living space under the ice, allowing the mink to make extended hunting forays, protected from the weather and its enemies.

The smallest weasels seem better adapted to the lake's surroundings. The ermine or stoat is probably most common in the woods, but the magnificent long-tailed weasel frequents open country. Some ten or twenty years ago, it was not unusual to see this lithe and serpentine creature popping in an out of gopher holes, or racing about with its tail straight up. During early spring, after snow cover is gone, the still-white weasels are plainly visible, but the long-tailed is rarely reported now. The species has declined elsewhere in Canada too and has been added to the official list of Threatened Animals.

The least weasel, North America's tiniest predator, is perhaps more common than the few observations indicate. In winter, it escapes detection with perfect camouflage. Even its stubby tail is all-white and lacks the black tip of the ermine and the long-tailed weasel. Of course, on bare ground, the tiny hunter is quite conspicuous and you may spot it running across a country road. Racing up the drifts on the other side, it will be gone in an instant, swallowed by the snow.

In summer, least weasels are seldom seen. One scurried over wet ground at the author's feet, evidently in search of a dry hiding place. Hardly larger than a mouse, it investigated shoreline wrack until it found shelter in a clump of reeds. It was probably a young of the year flushed out of its territory by waves that were inundating the shore, pushed by high winds.

Weasels are killed on sight by foxes and coyotes, as well as by hawks and owls, but they are not always eaten. Occasionally, the dried-up carcass of a least weasel may be found lying in the grass, showing tiny holes in the skin, punctured by talons.

It is a fact of life that big predators kill little predators, but collectively, none of these carnivores could exist were it not for the lowly rodents.

Chapter 21

FROM MICE TO MOOSE

Engrossed in their high-strung activities, meadow voles or "field mice" usually remain hidden from view under dense grass or deep snow. Insulated from cold, the prolific voles breed even in winter, especially if early fall snow covered unharvested grain, providing the rodents with unlimited food. In late March or April, when meltwaters flood their burrows, the plump, little creatures are forced out and run about erratically in search of dry ground. The country may swarm with them, and many are swept down by run-off water to lower ground, where the predators have a feast.

The common meadow vole has a blunt head and stubby tail. In brushy habitats, there are red-backed voles, whose chestnut-brown, glossy coat lends them some appeal to the human eye. But, as small mammals go, voles are not at all cute. By contrast, the deer mouse is a real dandy, a character straight out of Walt Disney's world of photogenic charmers. It is reddish-brown or warm-grey on its back and white underneath, the two colours sharply defined and highly contrasting. Even the long, tapering tail is bi-colour. The mouse's eyes and ears are large and ever-alert, as are the vibrissae bristling on the pointed nose, scanning for signals from friend or foe. This mouse is usually the critter that rattles in the dry underbrush at night, or that rummages through a camper's food supplies. If caught in the beam of a flashlight, it may stare back at you fearlessly, undisturbed by the glare as long as you remain still.

Here and there, concentrated around farms and granaries, are a few house mice, also equipped with large ears and a long tail, but lacking the colourful attractiveness of deer mice. House mice can be quite bold and run about behind your back while you sit quietly against a barn in the field, enjoying some springtime birding. These mice may even accept hand-outs and "share" your sandwich.

For the sake of completeness, a list of small mammals occurring at the lake should include several other mice and voles, as well as the tiny, insectivorous shrews, some of which weigh less than five grams. However, field identification of these dwarfs is all but impossible except to the specialist. Therefore, we shall skip to the next group of rodents, which are larger in size and easier to recognize.

A popular pet is the red squirrel, which occurs all across Canada wherever there is a stand of spruce trees, but it is practically absent from the deciduous woods around the lake. A great surprise is the local occurrence of flying squirrels! In 1988, soon after large nest boxes were placed around Lister Lake, in the hope of attracting cavity-nesting ducks or owls, half a dozen boxes were adopted by these nocturnal squirrels, that are common in all wooded regions of Alberta but very seldom seen by people.

Also numerous in its habitat of choice but rarely observed, is the pocket gopher, a subterranean dweller that leaves the familiar "mole" heaps of loose soil in pastures and meadows. Quite a bit smaller than the ubiquitous ground squirrel, which is popularly

Up close, the deer mouse is an attractive little creature, quite unlike the ''field mouse'' or meadow vole which has a stubby tail and small ears.

but mistakenly called "gopher", the real pocket gopher is a rotund, thickset fellow that shuns the sunlight and spends nearly its entire life in darkness. Unlike ground squirrels, it does not hibernate and remains active all winter. Its earth-lined tunnels lie like cables on the grass when the snow cover melts. Though very seldom seen by people, pocket gophers are often caught by hawks and owls. Sometimes they leave a dead but uneaten gopher on the ground or on a post, giving us a chance to examine the long, curved front claws, the prominent incisor teeth, and the storage pouch of loose skin on the side of the face from which it derives its name.

Because of the unsightly borrow mounds as well as the destruction of vegetables and ornamental flowers, cottagers have reason to hate the "moles". Farmers dread the lines of "mole heaps" invading freshly seeded fields, regardless of the fact that these animals are nature's little ploughmen that improve the quality of the soil by burying stubble and eating noxious weeds such as dandelions. Control efforts usually come down to the distribution of strychnine-treated grain, which is also used to kill ground squirrels. The latter are even more detested by the farmer, especially if they get into the crops. A single "gopher" pouch has been found to contain 260 kernels of wheat!

The use of poisoned grain may have fatal consequences for other animals as well, including the gopher's natural predators. Instead of strychnine, some farmers around the lake prefer to use a .22 rifle. Gopher shooting is considered a worthwhile Sunday entertainment in rural Alberta. Since ground squirrels are diurnal and often stand bold upright like picket stakes on the pastures, they make easy targets.

However, to the visiting naturalist, ground squirrels are a harmless and even comical sight, especially appreciated during early spring when birds are few and far between. The squirrels emerge from hibernation during March, even if deep snow still covers the ground. In fact, the deeper the snow cover is

during winter, the more squirrels may survive and the earlier they emerge, since the frost does not penetrate the ground as deeply as during a snow-less winter.

The distribution of ground squirrels is quite patchy, perhaps dependent upon soil conditions and control efforts. But during summer, lone squirrels can be encountered anywhere, apparently dispersing and in search of new territories. Without burrows for a quick escape, these travelling squirrels are at the mercy of predators, but enough survive to establish new colonies. There are now ground squirrels on the Lister Lake dam, which was built on wet ground, well away from existing squirrel habitat.

The familiar sand-coloured "gopher" of the Beaverhills area is called Richardson's ground squirrel, but in the wooded surroundings you may encounter two other kinds: the little thirteen-lined ground squirrel and the Franklin's ground squirrel. Both prefer brushy habitat and are seen infrequently, although summer dispersal may find them far from cover. A lone Franklin's was found hiding by a fence post on a wide-open section of west shore, a very unusual place for this pretty squirrel with its pearly-grey head and bushy tail.

Even more unusual is the local occurrence of black ground squirrels. The incidence of melanism in Richardson's ground squirrels is not mentioned in the handbooks, yet farmers see black "gophers" quite often, especially along the west shore. In June of 1982, there were four in one field. Unfortunately, this colony was decimated later, probably poisoned, and no blacks were reported thereafter. It stands to reason that the dark colour increases the risk of predation by hawks, or of being shot by people.

Similar risks are run by rodents that turn white before the snow flies. It is quite pathetic to see a white hare on bare ground. Nature's purpose of affording these much-hunted creatures extra camouflage and security only works if weather conditions cooperate. During the fall of 1987, when snow did not cover the

Varying hares or ''snowshoe rabbits'' have become quite numerous in the brushy woods in the Natural Area after grazing was stopped there.

ground until well into December, varying hares were quite numerous in the brush along the southeast shore. They were plainly visible to a great-horned owl perched on a snag. But catching a hare may not have been as easy as it seemed, since the animals stayed in the densest brush, protected from owl attacks by the tangled branches.

The larger of the two species of hares that occur at the lake, the big white-tailed jackrabbit, prefers open country and seldom takes shelter below bushes, confident in its ability to outrun and out-manoeuvre most of its enemies, including eagles and coyotes.

Close-up, the two hares look quite different. The smaller one, commonly called snowshoe rabbit, is a cute bunny with soft body contours and large, dark eyes. By contrast, the big jack glares at you balefully with light-coloured goat eyes. Its long legs and skinny ears give an impression of oddness and exaggerated proportions. When the animal is feeding in the grass, it hops awkwardly on folded rear legs, staying close to the ground and pressing its long ears over its back. But once the jack is on the move, the seemingly ugly is transformed into functional beauty. It is poetry in motion to see a jackrabbit running all-out, streaking over the ground or dancing in deliberate, stiff-legged bounds that send the snow flying.

During May, the jacks go courting and chase their bunnies through thick and thin, throwing caution to the wind. They zig-zag over the ground and through water in a single-minded urge to spend their passion and perpetuate their species that is so admirably adapted to live all year on the bleak open fields.

Rodents of a meeker disposition that make elaborate preparations for seasonal retirement, are the muskrat and the beaver. The winter houses of muskrats pop up all over the marsh during fall. Constructed of bulrushes and cattails, the biggest houses reach up to one metre above the lake's surface, allowing the inhabitants to come up out of the water to a dry bed hollowed out within the structure. The rats remain active all winter and make extended trips under the ice, expelling air

bubbles along the way that are replenished with oxygen from the water and utilized on the return trip. Here and there along their travel routes, the rats make so-called push-ups, small holes in the ice, covered over with plant material that prevents freezing. Food should be no problem for the rat, since emergent and aquatic vegetation is abundant in the marsh. Surprisingly, these vegetarians occasionally vary their diet with animal protein. One day, a rat was observed to eat 21 small fishes in succession. Each time, it would dive and emerge with a stickleback or minnow, evidently caught alive. The rat switched its flopping prey from its mouth to its front feet before munching it down, head first.

The aquatic mammal for which the lake and the nearby hills are named is of course the beaver, *Castor canadensis*. The hunt for its rich, dark-brown fur sparked the invasion of the northwest by trappers and traders a century ago. Nearly exterminated in central Alberta by the end of the 1800s, beavers have made a strong come-back and are now again very numerous in the hills where their favourite food, poplar, is the common tree species.

Beaverhills Lake itself, because of its shallow shoreline water, is not a good habitat for beaver. To survive winter, it must have access to a pond that does not freeze to the bottom and where the beavers can store their supply of branches in autumn. During island construction in Lister Lake, the draglines left deep water holes that are now the site of some large beaver lodges, built by the industrious animal engineers from sticks, earth and stones. In adjacent poplar groves, acres of trees have been cut, as evidenced by stumps and debris.

During spring, yearling beavers strike out on their own and can be encountered anywhere on the lake. One was sleeping on a floating iceberg off the tree-less west shore. In search of suitable territory, dispersing beavers swim or walk up any trickle of water running into the lake. Some end up in farm dug-outs or sloughs that offer no chance for a permanent settlement. Drying conditions may force these beavers to travel back to the lake over land, running the risk of being killed by coyotes or dogs.

Beavers are common in the major creeks, where they cut down the narrow strip of trees growing on the banks. Beaver trapping is a sad necessity for some landowners. Others call upon the heavy hand of government officers, who respond by dynamiting the dams, usually in winter, resulting in the callous destruction of the entire beaver family, as well as of fish that depend on the deep waterhole by the dam to make it through winter

The group of animals that completes this chapter seems to be universally loved as it creates no problems for people, at least not around the lake. This unusual group is comprised of three species of hoofed mammals or ungulates. Each of them would merit a detailed description, but a mere mention will have to do for this book. They are the white-

tailed deer, the mule deer and the moose, in order of frequency of occurrence.

For most of the year, deer stay close to trees and bushes, but they may be seen on open fields and in the marshes. It can be quite startling to surprise a large-antlered buck in the cattails, fleeing in great splashing bounds. Equally startling in another sense is the sight of a moose, gingerly stepping over barbed wire fences and loping across an open pasture. The grotesque native seems out of place among the cows and steers that have superseded the herds of bison, which ranged the lake shore before *Homo sapiens* asserted his presumed dominion over the land, the mammals and the birds.

Let us hope that his future stewardship will not be motivated by greed, but by wisdom, responsibility and tolerance, for the benefit of all wild creatures as well as their watchers.

Photos: Dick Dekker

Rising in Elk Island Park, Ross Creek enters the lake in the northwest. On one side, open pasture stretches for miles, on the other cultivation extends to the reedy shore, farther than anywhere else along the lake.

125

Chapter 22

LAND MANAGEMENT AND THE "BOVINE BULLDOZER."

After lake levels dropped in the 1920s, the margin that fell dry and extended beyond the surveyed boundaries of private lands, became the property of the government. The new "crown lands" were promptly leased to surrounding farmers for haying or grazing, but some parcels have since been deeded to landowners who had riparian rights of access to the lakeshore. It is to be hoped that this process can be halted and perhaps reversed. If the lake is to be managed and maintained for wildlife, it seems advisable that the shoreline remains under government control. At present, circa 4400 hectares of pasture around the lake add up to the largest chunk of crown land in central Alberta outside Elk Island National Park. This government-owned real estate, so close to a major population centre, will gain in importance for the recreational needs of naturalists and waterfowl hunters. The interests and viewpoints of these groups do not always coincide with those of private landowners, and conflicts easily arise concerning access.

A long-standing complaint of hunters has been that cattle are allowed to crop the pastures so short that ducks are forced to lay their eggs in poor nest cover, or to locate far from shore which increases the risks for ducklings of succumbing to predators or accidents on their long trek to water. It presents a clear case of competition, with farmers and the Public Lands Division on one side, hunters and the Fish and Wildlife Division on the other.

Of late years, much improvement has been

achieved in the cooperation between various agencies. To prevent overgrazing on crown lands, government inspectors have introduced cattle quotas and a system of pasture rotation. New fence lines were built to split the largest holdings into sections that can be grazed alternately. Of course, a cardinal factor that determines a pasture's grazing capacity is precipitation, which is out of anybody's control. What may be a realistic quota for an average season, is too generous during a drought, or too low for a wet summer. Sentient management still depends on responsible farmers who understand the interrelationship of cows, duck habitat and the vagaries of the weather.

In a bold move to improve communication between land-users and government officials, and to create long-range guidelines for the management of the crown lands around the lake, the Alberta Resource Evaluation and Planning Division drafted the so-called Integrated Land Use Plan. It divides the area into zones with different management priorities: wildlife, agriculture and recreation. These primary designations will be the major concerns in future management decisions affecting a particular piece of land. The Plan received support from land-users, naturalists and hunters, and it has resulted in some positive habitat protection. For instance, grazing was stopped in the wildlife zone around Lister Lake. However, a controversial management method is so-called pasture enhancement in areas zoned for agriculture. It involves the tearing up of natural sod and the

Landowners who have kept some native grassland intact are rewarded each spring with the purple flowers of prairie crocus. Buffalo bean (at left) and wild strawberry are common on most pastures.

reseeding of the area to tame grasses or grains that are more desirable from the farmer's point of view. As part of the improvements, bumps are levelled, holes filled, boulders removed or buried, and shrubs and trees destroyed. Most of the cost of the program is covered by government grants, with the lion's share ending up with the heavy-equipment operators.

Of late years, an increasing amount of energy has been expended on the clearing of pastures that were invaded by poplar and willow. This succession of grassland to woods is a natural sequence of events that can only be halted by fire or flood. The succession is retarded by a dense grass mat that gives seedlings little chance of getting established, and it is accelerated if the pasture is damaged by scraping and digging of animals. Bare ground is receptive to seeds.

In 1974, after the lake rose about one metre, much of the crown lands along the north side of the lake were inundated for two or three seasons, which killed all vegetation. When the land fell dry, poplar and willow seeds blew in from nearby woodlots, falling on fertile soils. Subsequently, a forest of shoots developed, much to the chagrin of the local lease-holder, who was neither permitted to burn the pasture, nor to use herbicides or a brush-cutter. Finally, in 1987, after the brush had grown into trees, the government habitat managers responded by sending in bulldozers, which scraped the sod and heaped the brush in ugly rows. This drastic method did not remove tree roots, which soon sent up shoots with a vengeance. In the spring of 1988, the pasture was treated with herbicides that killed the young growth of poplar and willow, as well as shrubs and broad-leaved plants.

A case can be made for the notion that the lake and the surrounding crown lands should not be managed at all. Let nature take its course, some say. It is an excellent idea for land where no grazing is tolerated, such as the marshy and wooded parcels east and west of Lister Lake. Plants and trees are developing undisturbed, creating a richer environment for mammals and birds in a natural process that will be more rewarding over time. However, there is good reason why other areas should not be allowed to succeed from grassland to woods if they are trampled and ruined by cattle. Moreover, it is of importance to retain pastures which are now open and without trees, since it is the required habitat for staging geese, shorebirds, some passerines and falcons.

During the past twenty years, formerly open points on the lake's north and northeast have grown up into brush, and as a result became less attractive to resting geese during the migration seasons. It would have been easy to prevent the loss of these traditional staging areas by burning the points during early spring when the soil is still frozen or wet. Formerly renowned Francis Point on the south side of the lake, where William Rowan and Robert Lister studied and collected birds in the first half of this century, was put to the torch each spring in a deliberate strategy used by most farmers to keep out brush and rejuvenate the grass. Francis Point is now overgrown with poplar and of no value for open-country birds.

If it can be managed with understanding as well as precision, periodic burning is perhaps the best method for retaining the open pastures. But a more subtle method is heavy grazing. The cow is man's cheapest bulldozer. It eats and tramples tree shoots along with the grass, and its manure, no matter how unsightly, fertilizes the soil.

On shoreline habitat, cattle have a serious negative impact. The heavy beasts scar the soft ground near the water, destroying nesting cover for ducks, and perhaps even crushing eggs and chicks. However, by uprooting the shallows and destroying bulrushes and cattails, cattle maintain the mudflats, which can be of service to shorebirds during years when open habitat is scarce. For instance, during the mid 1980s, after a decade of rather stable water levels had allowed the expansion of reeds along the entire west and north shore, the only significant shorebird habitat was located on two overgrazed pastures where cattle had kept a stretch of shore free of cattails and bulrush. Like other animals, cows are creatures of habit; year after year the local herd had come to the same spots to drink.

The grass on much of these pastures was always quite short, which made the locality a traditional feeding ground for thousands of migrating sandpipers and plovers that shun long grass. The trampled shore was used for resting and preening by ducks as well as shorebirds. They seek out an open, unobstructed environment since it lessens the risk of being caught in surprise attacks by their mortal enemy, the peregrine falcon. During summer, the reed-free areas were preferred by nesting avocets and killdeers, as well as by loafing gulls and geese, secure in the instinctive knowledge that the approach of any predator could be spotted quickly.

To some conservationists, the "bovine bulldozer" will no doubt remain a controversial presence around Beaverhills Lake, but it must be remembered that the dominant herbivore of the area used to be the cow's ancestor, the bison. It is an eye-opener to observe the impact that the great beasts have today in the Mackenzie Bison Range (NWT). There, the pitted ground and ruined cattails along large and shallow Falaise Lake resemble the worst of Beaverhills! On the prairies, the shaggy humpback's place has now been taken by the Hereford. Its effect on wildlife habitat is complex. Cattle should neither be excluded from all of the shoreline, nor have unlimited access. Perhaps, it would be best to establish a variable policy of no grazing at all in some areas and heavy or moderate utilization in others, ensuring a diversity of edge habitat and ground cover in which all kinds of birds and birdwatchers should find their particular niche.

At Francis Point, a well-marked gravel trail leads to the observation blind on the south shore.   Photo: Dick Dekker

Chapter 23

FROM MUDFLAT TO PASTURE

According to the accounts of early travellers who saw Beaverhills Lake in pre-settlement days, the surroundings were semi-open with vast expanses of "buffalo wool", a fine grass that was "as good as grain for a tired horse." Not much, if any, virgin sod remains today. Here and there, a small corner of dryland may have escaped the plow or has returned to a near-natural condition, with blue grama and fescue grasses and the purple blooms of prairie crocus in early spring.

Before turning the sod, the early settlers started fires to burn off brush and copses of trees. As reported in "Tales of Tofield," at least one early resident lamented the destruction of the Manitoba maples that were apparently indigenous to the area, at the western edge of their range in Canada. At present, a few maples grow in woodlots near the lake, but these trees were probably planted around former homesteads.

Trees have been allowed to make a big comeback in the pastures along the east and south shore, where they occupy mostly sandy soils. Among the poplar and willow stand a few white birch and alder, while the odd spruce has taken hold, its seed blown in from the nearest mature trees, perhaps more than a kilometre away. The understory in these woods is almost non-existent, grazed and trampled by the "bovine bulldozers" except in a section around Lister Lake where the government has established a no-grazing wildlife zone.

A little farther inland, on better soils near the farms or around wet spots in grain fields, are a few woodlots that hint at the area's original wealth of berry bushes and other shrubs. Saskatoon and chokecherry compete for space with snowberry and hazelnut. Wild rose, raspberry and a host of herbs crowd roadside ditches unless they are killed by the poisonous effects of herbicides, which turn the fields into monocultures of barley, oats, wheat or canola.

Fortunately, the plant community on the periphery of the lake has neither been destroyed nor changed much by man or beast. Along the shore, emergent vegetation and pioneering plant species succeed each other in a dynamic interplay that is ruled by water levels. Not coveted by man and safe from fires or the plow, the cattail grows year after year into a massive plant that sends its weird flower head into the air, high above the waterline. The dark-brown cylinder is said to resemble a cat's appendage, but in fact it looks more like a cigar. It hides innumerable tiny flowers that ripen into white seed fluff that escapes on autumn winds like puffs of smoke.

Also the common bulrush carries flower heads that produce numerous seeds, although, like the cattail, the plant spreads rapidly by sending out rootstocks across the lake bottom. The round, whip-like stalk is hollow, which makes underwater parts buoyant and allows air to reach the roots.

Less common than the cattail and the bulrush, but perhaps the most spectacular, is the two-metre high phragmites, that waves its plumes in dense stands. It is the common reed of Europe and eastern Canada. However, at Beaverhills Lake it grows only in two small stands on Dekker Island and near a pond along the east shore. In Saskatchewan and Manitoba, it is called yellow cane or giant reed grass. Like all true grasses, the phragmites' stem is hollow and segmented. Using a sharp knife, one can make a flute by cutting off a single segment and slicing a small slit into the thin wall.

The first vascular plant that colonizes newly-formed mudflats is the spectacular marsh ragwort. It begins to grow early in April, sprouting inch-thick hollow stems that carry a mass of distinctively hairy, bright yellow flowers. They mature into white fluff that transports the seeds in June. During summer, the leafy but fragile plant collapses and dies, to be replaced by another non-permanent pioneer species, the foxtail barley, familiar to all. In August, its bristle-like plumes turn silver and gold. They transform the mudflats into a luminous sea, waving in the wind. The young shoots of the foxtail make excellent fodder for cattle, but the mature, prickly awns are capable of causing severe inflammation of the animal's mouth. The base of the awns contains tiny seeds that are the origin of the plump and nutritious barley grain, developed by agronomists, that now grows commonly on the farm fields around the lake.

If the lake level stays low and the mudflats dry out, the foxtail barley will be replaced by small rushes, sedges, and grasses of many species that are very difficult to identify, even for the specialist. The three kinds of grass-like plants can be told apart by anyone, since their stems are either round, triangular, or segmented. This can be remembered easily with a catchy phrase: rushes are round, sedges have edges, and grasses make joints.

On higher ground away from the lake, the open grasslands feature a few species of flowering plants that locally occur in great numbers, their colours dominating the landscape and varying with the progress of the season. The showy buffalo bean is one of the earliest of native plants, but before its golden flowers can develop into the poisonous pods, an introduced weed, the ubiquitous dandelion, turns the meadows yellow. By early summer, another introduced weed, white clover, spreads sweet-smelling carpets that out-compete the native plants such as silverweed and blue-eyed grass over much of the short-grazed terrain.

Heavy rains stimulate the growth of grass. Fresh greens prevail until summer when common yarrow sends up armies of white florets, that contrast prettily with the yellow clusters of goldenrod and the starry-flowered blue lettuce. Tall and spindly, yellow sow-thistles wave in the wind over ripening grasses that form a multifarious background in olive-greens, tan and gold.

Here and there is a plant of special attraction that invites touch, taste or smell; the sticky gumweed, the delicious wild strawberry, or the fragrant mint and pasture sage. Others are overlooked and stepped upon, although they are noteworthy in their own way, such as the ground-hugging pussy-toes or everlasting, that can survive on very poor soil severely overgrazed by cattle.

Grazing pressure, next to precipitation and soil quality, is a decisive factor affecting plant composition and the height of grass. Across fence lines that divide one pasture from the other, there may be a striking difference in appearance. Heavy grazing may leave the ground-cover as short as a golf course. Eventually the grass may disappear altogether, to be replaced by small dandelions and other tolerant weeds. However, on many of the larger pastures around the lake, the effects of grazing are spotty. Evidently, cattle frequent some areas more than others. Very bald spots alternate with rank vegetation that is perhaps less palatable to bovine taste buds.

Comparative differences between well-used pastures and areas that have recently been excluded from grazing, where the vegetation is allowed to develop naturally, should become more interesting over time, to the botanist as well as the birdwatcher.

Chapter 24

WATERFOWL HUNTING AND MANAGEMENT

In the jargon of natural resource departments, hunting is a consumptive recreational use of wildlife, whereas birdwatching is a non-consumptive use. The term "use" implies the possibility of overuse and hence the need for management and legislation. Thusfar, lake-side birders have not been targeted for government regulation and restrictions other than the prohibition of motorized boats. With the exception of a no-trespassing zone around Pelican Island during the breeding season, visitors can go where and when they please. Such freedom can no longer be taken for granted elsewhere. At many birding hotspots near major population centres access is limited in order to prevent damage to fragile environments and to keep disturbance of breeding and staging birds to a minimum.

The notion that birdwatchers are non-consumptive may be true in the most literal sense that they do not eat their birds, but the binoculars-and-camera crowd is not quite harmless. Unintentionally and unwittingly, casualties occur. Visits to and photography of nests can have negative consequences for eggs and chicks that may be abandoned by parent birds and perish in cold weather, or fall prey to opportunistic predators. So-called hands-on birding experiences like netting and banding are not without hazards either, nor are scientific surveys of nesting colonial birds such as gulls, pelicans and herons. Even "viewing" has an effect on the subject of our one-sided love affair. Dutch ornithologists have expressed concern about the metabolic costs to wintering geese that are frequently disturbed on resting and feeding grounds by busloads of gawkers. In Britain, Holland and Sweden, where birdwatching has become phenomenally popular, the location of rare nesting birds is kept secret until the young have fledged. Such voluntary restrictions have become commonplace in the birdwatching community, which recognizes that it is non-consumptive only because its members do not intentionally seek to kill what they love, which is the hunter's paradox.

The need to protect birds from the gun became overdue around the turn of the century, when geese, ducks, waders, herons, and terns had been brought to near-extinction by year-round market hunting for meat and feathers.

The first legislated sanctuary for the protection of birds in North America was established by the State of California at Lake Merritt in 1870, two years before the creation of Yellowstone National Park. The first federal initiative in Canada for a bird refuge occurred in Saskatchewan in 1887, when land at the northern end of Last Mountain Lake was withdrawn from settlement, although the action was largely forgotten and the sanctuary was not granted a name until 1917.

Bird protection gained momentum in the 1880s with the formation of the American Ornithologists' Union and the Audubon Society. In 1903, President Theodore Roosevelt initiated a system of American National Wildlife Refuges with the designation of Pelican Island Refuge, located in the Indian River in Florida. The following year, Jack Miner established his private Canada Goose sanctuary in Kingsville, Ontario.

But outside these sanctuaries, the hunting of waterfowl was unrestricted until after 1916, when the North American Migratory Bird Treaty was signed by the U.S.A. and Great Britain on behalf of its Dominion of Canada. The government regulations not only limited the locations where waterfowl could be shot, but specified the times of year, the methods used and the number of birds a hunter was allowed to have in possession. Moreover, the new laws prohibited the sale of wild meat, effectively ending the greedy and unscrupulous practices of those who had slaughtered wildlife for monetary gain.

In Canada, hunting of migratory game birds has remained under federal jurisdiction although seasons and bag limits are set in cooperation with provincial authorities. In recognition of its potential for recreational hunting, Beaverhills Lake was declared a Public Shooting Ground in 1925, but a no-hunting zone of half a mile was enforced around the periphery to avoid disturbance to resting waterfowl. The protection is lifted on the first of November, so as to urge the migrants on their way south before winter sets in. Subject to yearly review, the shooting season at the lake usually begins in the second week of September and lasts until December.

This arrangement is generally approved and respected by local hunters, who understand that migrating geese need freedom from persecution on their loafing areas. If shot at on the lake, the birds leave and do not come back. The time to hunt is during morning and evening when the flocks visit inland stubble fields to feed. Hidden from view in willow blinds or shallow pits concealed with straw, surrounded by decoys, the hunter waits in the cold for a chance to fire his shotgun at low-flying birds. Such field-shoots used to be common practice. Of course, there are novices who cannot be bothered to make the classical preparations and instead crouch by a fence, pass-shooting whatever happens to come their way. Tempted by the many geese and ducks on the lake, some impatient rogues sneak close to the shore and disturb the resting flocks. Prior to 1973, illegal shooting by people hiding in the reedbeds along the north shore was a frequent irritant to responsible hunters and birdwatchers alike. In 1967, a new provincial wildlife director, Stuart Smith, made a bold move to enlarge the protective zone and to facilitate policing of the area by extending the shooting prohibition to the nearest county roads around the lake. Along the west and north, the no-shooting corridor became more than a mile wide and included fields that had been popular for pit-shoots. The measure was applauded tacitly by the Edmonton Bird Club, but protested vociferously by local landowners and hunters. The government quickly capitulated and restored the original half-mile zone, which remained in effect around the entire lake until 1973 when another persistent problem came to a head: waterfowl damage to unharvested crops.

To get away from its expensive policy of compensating farmers for grain lost to raiding geese and ducks, the Alberta Fish and Wildlife Division, in cooperation with the Canadian Wildlife Service, set up an experimental bait-station along the south shore to lure the birds away from the fields. Financed by a surcharge on hunting licences, a local farmer was contracted to dump a daily truck-load of grain in a fenced plot on the shore. The program proved very effective in keeping ducks off the land, to the chagrin of hunters who claimed that their field-shoots had been ruined and that the ducks were being fattened for the benefit of hunters south of the border. To placate these justified concerns, the government lifted the no-shooting regulation along the north half of the lake, opening the shore and the water to hunting, a trade-off that has remained in place until today, despite initial criticisms from old-timers who claimed that the open season on the north shore would result in the early departure of geese. The Wildlife Division defended its decision on two accounts. Firstly, it wanted to create more opportunities for city people to hunt on public lands. And secondly, if the geese were scared off the north half of the lake, they would simply move to the south end where the protective zone remained in effect. This shift did not happen for the simple reason that the wooded south shore is not as attractive to resting geese as the open pastures in the north.

Migrating geese continue to pass over the lake, in spring as well as fall, but they seldom come down during fall, or only very briefly. On October 13, 1988, five thousand snow geese descended along the north shores. White-fronted geese had been touching down flock after flock since August. But nearly all of these birds left again soon, perhaps moving on to Sullivan Lake in south-central Alberta, where their numbers often exceed forty thousand, and where both species linger for weeks each fall.

Very traditional in their choice of staging points, geese prefer wide-open shores and demand absolute freedom from human disturbance. Suitable conditions prevail at Sullivan but not at Beaverhills Lake, unless it is managed to its maximum potential for attracting staging geese. Suitable open habitat is limited to a few points along the north half, especially the large peninsula adjacent to Pelican Island. Here, the vegetation should be kept short by heavy grazing, haying and/or periodic burning, while access for people must be prohibited from August to November. At present, this section of shore is popular with hunters who like to take a canoe or punt into the hidden bays and reedbeds, accompanied by a well-trained dog. These solitude-loving wildfowlers have profited from the opening of the north half of the lake, which they usually have to themselves. For, ironically, there are fewer hunters now than there used to be. In Alberta, the sale of hunting licences has dropped from 80,000 in 1978 to less than 40,000 in 1986. This decline in the sport's popularity may reflect the rising cost of licences and equipment, as well as the decreasing opportunities. In the years when waterfowl numbers plunged across the prairie provinces, bag limits were reduced drastically, down to five ducks a day in 1988, including only one pintail and one canvasback. However, such species-specific bag limits may only be cosmetic and uneffective, since many hunters have trouble distinguishing one duck from another, either in the air or in the hand. In a five-year investigation of hunter performance in the field, biologists of the

Canadian Wildlife Service found that 40% of Alberta hunters could not identify the ducks they had killed. Spying on the shooters from a blind, or by pretending to be hunters themselves, the biologists also monitored illegal behaviour, such as shooting at protected species, discarding birds and failing to retrieve cripples. Fifteen percent of Alberta hunters were seen to violate regulations. Marsh hunters often fired at gulls, shorebirds, swans, or grebes. Over-water hunters who had no dogs seldom made an attempt to find cripples unless downed birds were within easy reach. Pass-shooters, either in marsh or field, frequently shot at birds beyond effective killing range and failed to retrieve a large number of wounded birds. Crippling losses were estimated at 20-45%. Decoy hunters were generally better equipped to retrieve cripples and they experienced higher success rates, which tempted them to overbag and discard the surplus.

Overall, hunter success rates were very low, with more than 70% of Alberta shooters unable to take even one duck or goose a day. Less than 6% attained the daily bag limit. The C.W.S. study also reported that many of the violations were committed by novice hunters, who lacked the experience and equipment to hunt properly and who caused high crippling losses, which points to the need for instruction and guidance before people gain the right to use a gun.

Except monitoring losses on the waterfowl balance sheet, government and private agencies are increasingly focussing on the credit side: how to boost duck production on the prairies in the face of accelerating losses of wetland habitat due to drainage and drought conditions. Ducks Unlimited Canada, initiated by American waterfowl hunters, has been active for half a century in the preservation and improvement of duck breeding habitat. In Alberta, D.U. biologists and engineers have combined their talents on more than 1300 sites. Beaverhills Lake has received attention since 1971 and to date five local projects have been completed, designed to restore or regulate the natural flow of water through the use of control structures in creeks and run-off channels. The southeast corner has three D.U. projects: Lister Lake was stabilized by a weir and enhanced with islands; the Kallal and Lawrence projects use water from Amisk Creek to create marsh habitats. In the northwest corner of the lake, the so-called Ronnie North and Ronnie South projects preserve water that would normally be lost early in spring as run-off to the main lake. Two projects along the east shore, lake "C" and "Inland", inundate and revitalize natural depressions that were formerly bays of the lake when its level was much higher.

All projects were preceded by negotiations with landowners, some of whom gained by increased hay production, especially in the Lawrence and Ronnie North projects, where water levels are lowered in summer for that purpose. Government departments are involved as well. In 1982, the Alberta Fish and Wildlife Division entered into a major agreement with Ducks Unlimited. One partner secures the land and the other contributes the money for the development of the twenty most important wetlands in the province, including Beaverhills Lake. Titled "Wetlands for Tomorrow", the agreement bodes well for consumptive users of the waterfowl resource as well as for non-consumptive users, especially since the water management schemes not only benefit ducks and geese, but a host of other birds and mammals.

To cut down on crop losses to raiding waterfowl, government wildlife agencies operate a "duck lure" bait station on the lake's south shore.

EPILOGUE

Since 1991, when *Prairie Water* was first published, major changes have taken place around the lake. While its popularity has increased by leaps and bounds, its water levels have plummeted. In the fall of 1995, as viewed from the same vantage point as in the photos on page 20, the main lake was literally out of sight. However, long-time residents recall that it dropped even lower during the early 1950s. Fortunately, in 1997 snow run-off and rain brought the lake back up to a respectable level. In the meantime, it was fascinating to observe the changes the down-cycle brought to the ecosystem.

As the shoreline retreated, cattails and bulrushes withered. The mudflats sprouted masses of seedlings, but they too died after the water failed to come back up. Homeless, some blackbirds built their nests in ragwort, which grew to giant proportions. Other species took advantage of open ground. In 1993, the rare piping plover returned to nest successfully for the first time in many years. In 1996 there were six pairs, but much of their habitat was subsequently invaded by foxtails. One or two seasons later, it turned into a jungle of chest-high reed grass, thistles and other hardy weeds. A walk along the shore felt like an obstacle course, particularly where cattle had pock-marked the soft ground.

Depending on rainfall and wind direction, the mudflats alternately expanded or shrank. They became a magnet for migrating shorebirds. In 1995, biologist Michael Barr of Ducks Unlimited organized the first-ever comprehensive shorebird census, covering the entire shoreline. On May 19 and 25, the participants, driving quad-buggies, obtained daily counts of just over 50,000. It qualified the lake for new distinctions. On May 28, 1996, it was designated as a part of the Western Hemisphere Shorebird Reserve Network, the fourth such site in Canada. In April 1997, the lake was identified as an "Important Bird Area of Global Significance", an initiative of BirdLife International.

The wide lake margins of 1995 and 1996 also favoured geese, especially during spring. Snow geese became more abundant than ever, reflecting their recovery on northern nesting grounds. However, during fall, their flocks remained minimal. Swans all but deserted the main lake because of the disappearance of pondweeds on which they feed. A few of the stately white birds still congregate in Lister Lake, where the water level has remained constant and cattails have survived, thanks to the weir.

The impact of dropping water levels was most severe on birds that feed on fish and shrimps, which became practically non-existent. Grebes, night herons and terns stopped breeding and went elsewhere. More than 300 pairs of pelicans and about 100 cormorants held on to their island, although they had to commute daily to other lakes, such as Hastings, to obtain food. However, on June 27, 1993, the author found the island abandoned; not a single pelican egg remained, and the nests of cormorants contained only dead chicks. Tracks indicated that coyotes had raided the breeding colony by crossing the shallow channel between island and shore.

As just about everywhere in Alberta, coyotes are ubiquitous. Along the west shore, they inter-bred with a small white and black collie, which set up a thriving dynasty of coy-dogs, most of them black. As well as preying on ground-nesting birds, coyotes kill a large range of mammalian prey, including any weasel they can catch. The beautiful long-tailed weasel has become scarce around the lake. Jack rabbits and skunks have also declined steeply. However, the demise of the latter may have other causes. Alberta biologists John Gunson and Ron Bjorge, who did an intensive study of skunks in the Tofield area in the 1970s, discovered that females den communally under buildings and barns. Up to 18 females were found denning together, accompanied by one male! Since many of the old-style wooden field granaries have been torn down, the skunks may have been forced out of house and home. Their survival is also adversely affected by cold winters when snow depth is insufficient to insulate the skunk's hibernation chamber. Winters with little snow have been typical in the recent past. The lack of spring run-off is the primary reason why water levels have dropped.

The lake's wildlife are affected by the vagaries and interaction of many factors, both natural and human-related. Over the last few years, some species have increased, such as the raven. Others have lost ground. One of these, unfortunately, is the sharp-tailed grouse. A local ban on hunting upland birds around the lake might bring them back in the future. (Speaking of hunting, the season on rails has been closed. But no-one was shooting them anyway.)

The Snow Goose Festival

A spectacular event that has become a new tradition at Tofield is the Snow Goose Festival, which takes place in late April. In 1998, its seventh year, more than 6000 people attended the event, which includes guided bus tours, hikes and wildlife displays. The festival was initiated by the Canadian Wildlife Service with the partici-pation of several other agencies, including the Beaverhill Lake Nature Centre, which now includes a budding museum. The dedicated work of many professionals and volunteers has enhanced the lake's appeal for increasing numbers of people. This can only lead to a consolidation of concern for its wild heritage.

CHECKLIST OF THE BIRDS OF BEAVERHILLS LAKE

The bird names and species sequence used in this book are those of the Revised Edition (1986) of *The Birds of Canada* (W. Earl Godfrey), which largely follows the American Ornithologists' Union's *Checklist of North American Birds* (Sixth Edition, 1983).

The following list pertains to all of the lake and its shores, as well as surrounding pastures, fields and woods, to the nearest country roads (see map). Up to fall 1990, 267 bird species have been recorded in this area. Species that the visitor can expect to see with a fair degree of certainty in the right habitat and at the right time of year, are marked with a dash (-). Open circles (○) indicate species that have often been reported but not each year and on an irregular and unpredictable basis. Species recorded less than half a dozen times since 1960 are marked with a solid symbol (●).

Birds known or assumed to be *breeding* in the checklist area may be present during all or part of spring, summer and fall. *Transients* occur during migration when they pass through on their seasonal journeys to and from northern breeding grounds. Species described as *vagrants* are considered to be out of normal range, such as long-billed curlews and ferruginous hawks, which breed well south of the Beaverhills region, but occasionally stray north. For species in the rare (solid dot) category, dates of reported sightings are given, when available, and the names of the observers are indicated by their initials (see list at right).

As discussed in the chapter titled "Rarities", the concept "rare bird" is subjective and relative, depending on opinions and birding habits. Smith's longspur and dunlin, reported quite often in open habitat, may never be seen by people who spend most of their time banding songbirds in the willows near the Beaverhill Bird Observatory.

The checklist is based on the published literature, field notes and opinions of a complement of birders.

Edgar Jones and Stefan Jungkind provided most of the information on woodland birds and some ducks; the author's data prevailed for shorebirds, raptors and some open-country passerines. An earlier version of the checklist was critically reviewed by Rainer Ebel. The 1997 updates are the product of seven years of meticulous record-keeping by Roy Fairweather. No doubt, the growing number of keen birders afield ensures that this checklist will be in need of further updates soon after publication. However, this list should provide a reference source for use far into the future. For current information on the lake's birds, see the annual reports and special publications of the Beaverhill Bird Observatory and other agencies.

The following people, represented by their initials, reported observations of birds considered rare and marked with the symbol ●: David Boag (DB), Dick Dekker (DD), Peter DeMulder (PDM), Peter Dunn (PD), Rainer Ebel (RE), Bob Gehlert (BG), Loran Goulden (LG), Cy Hampson (CH), Wes Hochachka (WH), Otto Hohn (OH), Harry Horton (HH), Edgar Jones (EJ), Stefan Jungkind (SJ), Richard Klauke (RK), Jim Lange (JL), Bill Lea (BL), H. MacGregor (HM), Chel Macdonald (CM), Martin McNicholl (MM), Dave Nadeau (DN), Mike Quinn (MQ), Jay Riddell (JR), W. Ray Salt (WRS), Ron Slagter (RS), Petra Stubbs (PS), Philip Taylor (PT), Terry Thormin (TT), Ken Trann (KT), Eric Tull (ET), Robert Turner (RT), Eric Wallace (EW), Chip Weseloh (CW), Alan Wiseley (AW).

Records marked RL are taken from Robert Lister's *The Birds and Birders of Beaverhills Lake*. The initials SS stand for Salt and Salt's *The Birds of Alberta*. The symbol + behind initials means "and other observers". Unless otherwise indicated, all reports involve the sighting of single birds.

● Red-throated Loon *Gavia stellata* (3 Oct. 1979, DD)
○ Common Loon *Gavia immer* (transient)

- Pied-billed Grebe *Podilymbus podiceps* (breeding)
- Horned Grebe *Podiceps auritus* (breeding)
○ Red-necked Grebe *Podiceps grisegena* (breeding)
- Eared Grebe *Podiceps nigricollis* (breeding)
- Western Grebe *Aechmophorus occidentalis* (breeding)
● Clark's Grebe *Aechmophorus clarkii* (15 Sept. 1990, EM)
- American White Pelican *Pelecanus erythrorhynchos* (breeding)
- Double-crested Cormorant *Phalacrocorax auritus* (breeding)
- American Bittern *Botaurus lentiginosus* (breeding)
- Great Blue Heron *Ardea herodias* (transient and summer resident; bred in 1970s)
● Great Egret *Casmerodius albus* (23 May 1976, DD; 16 May 1981, BG; 7 June 1987, PD+)
● Snowy Egret *Egretta thula* (June 1984, CM)
- Black-crowned Night-heron *Nycticorax nycticorax* (breeding)
- Tundra Swan *Cygnus columbianus* (transient; few summer records)
● Trumpeter Swan *Cygnus buccinator* (few fall records identified by neck bands; 3 Oct. 1988, RK; 9 Oct. 1989, DD, identified by calls)
- Greater White-fronted Goose *Anser albifrons* (transient)
- Snow Goose *Anser caerulescens* (transient; few summer records)
○ Ross's Goose *Anser rossini* (transient)
● Brant *Branta bernicla* (17 Oct. 1971; 8 May 1978; 13 Sept. 1978; pair, 24 Sept. 1978, DD)
- Canada Goose *Branta canadensis* (breeding)
● Wood Duck *Aix sponsa* (3 May 1980, RE; 29 May 1985, DD)
- Green-winged Teal *Anas crecca* (breeding)
○ American Black Duck *Anas rubripes* (transient)
- Mallard *Anas platyrhynchos* (breeding)
- Northern Pintail *Anas acuta* (breeding)
- Blue-winged Teal *Anas discors* (breeding)
- Cinnamon Teal *Anas cyanoptera* (breeding)
- Northern Shoveler *Anas clypeata* (breeding)
- Gadwall *Anas strepera* (breeding)
● Eurasian Wigeon *Anas penelope* (29 Apr. 1984, RK+; 24 Apr. 1986, PM; 5 Apr. 1987, JL+EW; 26 Sept. 1990, RK)
- American Wigeon *Anas americana* (breeding)
- Canvasback *Aythya valisineria* (breeding)
- Redhead *Aythya americana* (breeding)
○ Ring-necked Duck *Aythya collaris* (transient)
○ Greater Scaup *Aythya marila* (transient)
- Lesser Scaup *Aythya affinis* (breeding)

● Harlequin Duck *Histrionicus histrionicus* (one undated record, SS)
● Oldsquaw *Clangula hyemalis* (4 Oct. 1964, RL+RT; 29 Apr. 1984, SJ)
● Black Scoter *Melanitta nigra* (pair, 15 May 1982, EJ)
○ Surf Scoter *Melanitta perspicillata* (transient)
- White-winged Scoter *Melanitta fusca* (breeding)
- Common Goldeneye *Bucephala clangula* (breeding)
○ Barrow's Goldeneye *Bucephala islandica* (vagrant)
- Bufflehead *Bucephala albeola* (transient; possibly breeding)
○ Hooded Merganser *Lophodytes cucullatus* (transient; few summer records)
- Common Merganser *Mergus merganser* (transient)
- Red-breasted Merganser *Mergus serrator* (transient)
- Ruddy Duck *Oxyura jamaicensis* (breeding)
● Turkey Vulture *Cathartes aura* (2 Sept. 1924, RL; 3 May 1958; 18 July 1961, EJ; 14 May 1967; 26 Aug. 1967, DD)
○ Osprey *Pandion haliaetus* (transient)
- Bald Eagle *Haliaeetus leucocephalus* (transient)
- Northern Harrier *Circus cyaneus* (breeding)
- Sharp-shinned Hawk *Accipiter striatus* (breeding)
- Cooper's Hawk *Accipiter cooperii* (breeding)
○ Northern Goshawk *Accipiter gentilis* (transient)
○ Broad-winged Hawk *Buteo platypterus* (transient)
- Swainson's Hawk *Buteo swainsoni* (breeding)
- Red-tailed Hawk *Buteo jamaicensis* (breeding)
○ Ferruginous Hawk *Buteo regalis* (vagrant)
- Rough-legged Hawk *Buteo lagopus* (transient)
○ Golden Eagle *Aquila chrysaetos* (transient)
- American Kestrel *Falco sparverius* (transient)
- Merlin *Falco columbarius* (breeding)
- Peregrine Falcon *Falco peregrinus* (transient)
○ Gyrfalcon *Falco rusticolus* (transient)
○ Prairie Falcon *Falco mexicanus* (vagrant)
- Gray Partridge *Perdix perdix* (breeding and wintering)
- Ring-necked Pheasant *Phasianus colchicus* (captive-raised and released; possibly breeding)
- Ruffed Grouse *Bonasa umbellus* (breeding and wintering)
● Greater Prairie-Chicken *Tympanuchus cupido* (extirpated; last seen in 1932)
- Sharp-tailed Grouse *Tympanuchus phasianellus* (breeding and wintering)
- Yellow Rail *Coturnicops noveboracensis* (assumed breeding)
● Virginia Rail *Rallus limicola* (heard calling, 25-27 May 1984, EJ)
- Sora *Porzana carolina* (breeding)
- American Coot *Fulica americana* (breeding)
- Sandhill Crane *Grus canadensis* (transient)

- Whooping Crane *Grus americana* (flock of three, 19 Apr. 1957, DB+WRS; single bird, May 1959, CH+HM; flock of three, 1 Oct. 1966, DD)
- Black-bellied Plover *Pluvialis squatarola* (transient)
- American Golden Plover *Pluvialis dominica* (transient)
- Semipalmated Plover *Charadrius semipalmatus* (transient)
- Piping Plover *Charadrius melodus* (vagrant; possibly breeding in 1976-78 and again in 1993-96)
- Killdeer *Charadrius vociferus* (breeding)
- Black-necked Stilt *Himantopus mexicanus* (three nests in 1977, DD+CW; sight record in May 1978, RS)
- American Avocet *Recurvirostra americana* (breeding)
- Greater Yellowlegs *Tringa melanoleuca* (transient)
- Lesser Yellowlegs *Tringa flavipes* (transient; likely breeding)
- Solitary Sandpiper *Tringa solitaria* (transient)
- Willet *Catoptrophorus semipalmatus* (breeding)
- Wandering Tattler *Heteroscelus incanus* (four birds, 2 Sept. 1973, OH)
- Spotted Sandpiper *Actitis macularia* (breeding?)
- Upland Sandpiper *Bartramia longicauda* (transient)
- Whimbrel *Numenius phaeopus* (transient)
- Long-billed Curlew *Numenius americanus* (vagrant)
- Hudsonian Godwit *Limosa haemastica* (transient)
- Marbled Godwit *Limosa fedoa* (breeding)
- Ruddy Turnstone *Arenaria interpres* (transient)
- Surfbird *Aphriza virgata* (21 Sept. 1975, RK+)
- Red Knot *Calidris canutus* (transient)
- Sanderling *Calidris alba* (transient)
- Semipalmated Sandpiper *Calidris pusilla* (transient)
- Western Sandpiper *Calidris mauri* (transient)
- Least Sandpiper *Calidris minutilla* (transient)
- White-rumped Sandpiper *Calidris fuscicollis* (transient)
- Baird's Sandpiper *Calidris bairdii* (transient)
- Pectoral Sandpiper *Calidris melanotos* (transient)
- Sharp-tailed Sandpiper *Calidris acuminata* (transient)
- Dunlin *Calidris alpina* (transient)
- Stilt Sandpiper *Calidris himantopus* (transient)
- Buff-breasted Sandpiper *Tryngites subruficollis* (transient)
- Ruff *Philomachus pugnax* (female, 8 May 1978, DD)
- Short-billed Dowitcher *Limnodromus griseus* (transient)
- Long-billed Dowitcher *Limnodromus scolopaceus* (transient)
- Common Snipe *Gallinago gallinago* (breeding)
- Wilson's Phalarope *Phalaropus tricolor* (breeding)
- Red-necked Phalarope *Phalaropus lobatus* (transient)
- Red Phalarope *Phalaropus fulicaria* (transient)
- Parasitic Jaeger *Stercorarius parasiticus* (transient)
- Long-tailed Jaeger *Stercorarius longicaudus* (1924, RL; 8-12 Sept. 1977, ET+DD+)
- Franklin's Gull *Larus pipixcan* (breeding)
- Little Gull *Larus minutus* (14-18 Oct. 1986, RK+)
- Bonaparte's Gull *Larus philadelphis* (transient)
- Mew Gull *Larus canus* (8 Aug. 1988, DD)
- Ring-billed Gull *Larus delawarensis* (breeding)
- California Gull *Larus californicus* (breeding)
- Herring Gull *Larus argentatus* (transient)
- Iceland Gull *Larus glaucoides* (22 Oct. 1926, RL)
- Glaucous Gull *Larus hyperboreus* (transient)
- Black-legged Kittiwake *Rissa tridactyla* (15 May 1988, RE)
- Sabine's Gull *Xema sabini* (Sept. 1928, SS; flock of 30, 3 June 1972, PT+DD; 5 Sept. 1975, DD)
- Common Tern *Sterna hirundo* (breeding)
- Forster's Tern *Sterna forsteri* (breeding)
- Black Tern *Chlidonias niger* (breeding)
- Ancient Murrelet *Synthliboramphus antiquus* (2 Oct. 1983, DD)
- Rock Dove *Columba livia* (breeding and wintering)
- Mourning Dove *Zenaida macroura* (probably breeding)
- Black-billed Cuckoo *Coccyzus erythropthalmus* (breeding)
- Great Horned Owl *Bubo virginianus* (breeding and wintering)
- Snowy Owl *Nyctea scandiaca* (transient and wintering)
- Northern Hawk-Owl *Surnia ulula* (13 Oct. 1973, LG+DD)
- Burrowing Owl *Athene cunicularia* (30 May 1960, BG; 5 May 1968, PDM+RT; 1-20 May 1982, DD+RS; 7 May 1983, RS)
- Long-eared Owl *Asio otus* (breeding)
- Boreal Owl *Aegolius funereus* (collected, 1925, RL; found dead, 15 May 1983, EJ)
- Short-eared Owl *Asio flammeus* (breeding and wintering)
- Northern Saw-whet Owl *Aegolius acadicus* (transient; one breeding record in nestbox, RE+)
- Common Nighthawk *Chordeiles minor* (transient)
- Ruby-throated Hummingbird *Archilochus colubris* (vagrant)
- Belted Kingfisher *Ceryle alcyon* (14 Sept. 1974, 22 Aug. 1987, 12 May 1990, DD)
- Red-headed Woodpecker *Melanerpes erythrocephalus* (24 May 1987, JR+DN)

○ Yellow-bellied Sapsucker *Sphyrapicus varius* (transient)
- Downy Woodpecker *Picoides pubescens* (breeding and wintering)
○ Hairy Woodpecker *Picoides villosus* (vagrant)
- Northern Flicker *Colaptes auratus* (breeding?)
● Pileated Woodpecker *Dryocopus pileatus* (12 Sept. 1987; 30 Apr. 1988, EM)
○ Olive-sided Flycatcher *Contopus borealis* (transient)
- Western Wood-Pewee *Contopus sordidulus* (transient; possibly breeding)
○ Yellow-bellied Flycatcher *Empidonax flaviventris* (transient)
- Alder Flycatcher *Empidonax alnorum* (breeding)
○ Willow Flycatcher *Empidonax trailii* (transient)
- Least Flycatcher *Empidonax minimus* (breeding)
○ Eastern Phoebe *Sayornis phoebe* (breeding)
○ Say's Phoebe *Sayornis saya* (transient)
● Great Crested Flycatcher *Myiarchus crinitus* (heard on 12 July 1987, RE)
● Western Kingbird *Tyrannis verticalis* (7 June 1979, RK)
- Eastern Kingbird *Tyrannus tyrannus* (breeding)
- Horned Lark *Eremophila alpestris* (breeding)
- Purple Martin *Progne subis* (vagrant; breeds in Tofield)
- Tree Swallow *Tachycineta bicolor* (breeding)
○ Northern Rough-winged Swallow *Stelgidopteryx serripennis* (vagrant)
- Bank Swallow *Riparia riparia* (transient)
- Cliff Swallow *Hirundo pyrrhonota* (breeding under Amisk bridge)
- Barn Swallow *Hirundo rustica* (breeding)
- Blue Jay *Cyanocitta cristata* (transient)
- Black-billed Magpie *Pica pica* (breeding and wintering)
- American Crow *Corvus brachyrhynchos* (breeding)
○ Common Raven *Corvus corax* (breeding and wintering)
- Black-capped Chickadee *Parus atricapillus* (breeding and wintering)
○ Boreal Chickadee *Parus hudsonicus* (transient)
- Red-breasted Nuthatch *Sitta canadensis* (transient; breeding not confirmed)
○ Brown Creeper *Certhia americana* (transient)
- House Wren *Troglodytes aedon* (breeding)
○ Sedge Wren *Cistothorus platensis* (breeding, WH)
- Marsh Wren *Cistothorus palustris* (breeding)
○ Golden-crowned Kinglet *Regulus satrapa* (transient)
- Ruby-crowned Kinglet *Regulus calendula* (transient)
- Mountain Bluebird *Sialia currucoides* (breeding)
● Townsend's Solitaire *Myadestes townsendi* (23 Apr. 1967, RT+; four birds, 11 Apr. 1981, DD)

○ Veery *Catharus fuscescens* (breeding)
● Gray-cheeked Thrush *Catharus minimus* (two banding records, May 1985, 1986, EJ)
- Swainson's Thrush *Catharus ustulatus* (transient)
○ Hermit Thrush *Catharus guttatus* (transient; possibly breeding)
- American Robin *Turdus migratorius* (breeding)
○ Gray Catbird *Dumetella carolinensis* (transient; possibly breeding)
● Northern Mockingbird *Mimus polyglottos* (14 May 1975, DD)
○ Brown Thrasher *Toxostoma rufum* (breeding, EJ)
- American Pipit *Anthus spinoletta* (breeding)
- Sprague's Pipit *Anthus spragueii* (breeding)
○ Bohemian Waxwing *Bombycilla garrulus* (transient)
- Cedar Waxwing *Bombycilla cedrorum* (breeding)
○ Northern Shrike *Lanius excubitor* (transient)
○ Loggerhead Shrike *Lanius ludovicianus* (breeding)
- European Starling *Sturnus vulgaris* (breeding)
○ Solitary Vireo *Vireo solitarius* (transient)
- Warbling Vireo *Vireo gilvus* (breeding)
○ Philadelphia Vireo *Vireo philadelphicus* (transient)
- Red-eyed Vireo *Vireo olivaceus* (breeding)
- Tennessee Warbler *Vermivora peregrina* (transient)
- Orange-crowned Warbler *Vermivora celata* (transient)
- Yellow Warbler *Dendroica petechia* (breeding)
● Chestnut-sided Warbler *Dendroica pennsylvanica* (one banding record, 30 Aug. 1990, SJ)
- Magnolia Warbler *Dendroica magnolia* (transient)
○ Cape May Warbler *Dendroica tigrina* (transient)
● Black-throated Blue Warbler *Dendroica caerulescens* (4 Oct. 1928, SS)
- Yellow-rumped Warbler *Dendroica coronata* (transient)
○ Black-throated Green Warbler *Dendroica virens* (transient)
● Blackburnian Warbler *Dendroica fusca* (24 Aug. 1989, SJ)
- Palm Warbler *Dendroica palmarum* (transient)
○ Bay-breasted Warbler *Dendroica castanea* (transient)
○ Blackpoll Warbler *Dendroica striata* (transient)
○ Black-and-white Warbler *Mniotilta varia* (transient)
- American Redstart *Setophaga ruticilla* (transient; possibly breeding)
○ Ovenbird *Seiurus aurocapillus* (transient)
○ Northern Waterthrush *Seiurus noveboracensis* (transient)
○ Mourning Warbler *Oporornis philadelphia* (transient)
- Common Yellowthroat *Geothlypis trichas* (breeding)

- Wilson's Warbler *Wilsonia pusilla* (transient)
○ Canada Warbler *Wilsonia canadensis* (transient)
- Rose-breasted Grosbeak *Pheucticus ludovicianus* (transient; possibly breeding)
○ Western Tanager *Piranga ludoviciana* (transient)
● Indigo Bunting *Passerina cyanea* (24 May 1990, EJ)
- American Tree Sparrow *Spizella arborea* (transient)
- Chipping Sparrow *Spizella passerina* (breeding)
- Clay-coloured Sparrow *Spizella pallida* (breeding)
- Vesper Sparrow *Pooecetes gramineus* (breeding)
● Lark Bunting *Calamospiza melanocorys* (June 1966, BL; 15 May 1984, WH)
- Savannah Sparrow *Passerculus sandwichensis* (breeding)
○ Baird's Sparrow *Ammodramus bairdii* (breeding)
- LeConte's Sparrow *Ammodramus leconteii* (breeding)
- Sharp-tailed Sparrow *Ammodramus caudacutus* (breeding)
○ Fox Sparrow *Passerella iliaca* (transient)
- Song Sparrow *Melospiza melodia* (breeding)
- Lincoln's Sparrow *Melospiza lincolnii* (transient)
- Swamp Sparrow *Melospiza georgiana* (breeding)
- White-throated Sparrow *Zonotrichia albicollis* (transient)
- White-crowned Sparrow *Zonotrichia leucophrys* (transient)
○ Harris' Sparrow *Zonotrichia querula* (transient)
- Dark-eyed Junco *Junco hyemalis* (transient)
● McCown's Longspur *Calcarius mccownii* (12 Sept. 1972, MM)
- Lapland Longspur *Calcarius lapponicus* (transient)
○ Smith's Longspur *Calcarius pictus* (transient)
○ Chestnut-collared Longspur *Calcarius ornatus* (bred prior to 1970, DD)
- Snow Bunting *Plectrophenax nivalis* (transient)
- Bobolink *Dolichonyx oryzivorus* (breeding)
- Red-winged Blackbird *Agelaius phoeniceus* (breeding)
- Western Meadowlark *Sturnella neglecta* (breeding)
- Yellow-headed Blackbird *Xanthocephalus xanthocephalus* (breeding)
○ Rusty Blackbird *Euphagus carolinus* (transient)
- Brewer's Blackbird *Euphagus cyanocephalus* (breeding)
- Common Grackle *Quiscalus quiscula* (breeding)
- Brown-headed Cowbird *Molothrus ater* (breeding)
- Northern Oriole *Icterus galbula* (breeding)
- Purple Finch *Carpodacus purpureus* (transient)
○ Common Redpoll *Carduelis flammea* (transient and wintering)
○ Pine Siskin *Carduelis pinus* (transient)
- American Goldfinch *Carduelis tristis* (breeding)

○ Evening Grosbeak *Coccothraustes vespertinus* (transient)
- House Sparrow *Passer domesticus* (breeding and wintering)

Hypothetical records:
? Yellow-billed Loon *Gavia adamsii* (18 Oct. 1924, RL)
? Snowy Plover *Charadrius alexandrinus* (31 May 1975, DD+RS)
? Black Turnstone *Arenaria melanocephala* (23 May 1923, RL)
? Caspian Tern *Sterna caspia* (23 Sept. 1923, RL)
? Great Grey Owl *Strix nebulosa* (1925, RL)

Since 1991, there have been repeat sightings of 22 rarities: Red-throated Loon, Great Egret, Brant, Wood Duck, Eurasian Wigeon, Harlequin Duck, Turkey Vulture, Black-necked Stilt, Long-tailed Jaeger, Little Gull, Sabine's Gull, Great Gray Owl, Pileated Woodpecker, Great-crested Flycatcher, Western Kingbird, Townsend's Solitaire, Gray-cheeked Thursh, Northern Mockingbird, Chestnut-sided Warbler, Black-throated Blue Warbler, Lark Bunting. These updates were provided by Roy Fairweather, who has compiled a yearly checklist since 1991.

Since 1991, the following 19 new species have been recorded at the lake, bringing the grand total up to 291, including 4 hypothetical sightings. Species marked (s) are field sightings. All others were netted and banded by the Beaverhill Bird Observatory or by Edgar Jones. For further information and updates, contact Roy Fairweather and/or the Beaverhill Bird Observatory.

Caspian Tern *Sterna caspia* (s)
Arctic Tern *Sterna paradisaea* (s)
Barred Owl *Strix varia* (s)
Gray Jay *Perisoreus canadensis* (s)
White-breasted Nuthatch *Sitta carolinensis*
Winter Wren *Troglodytes troglodytes*
Wood Thrush *Hylocichla mustelina*
Varied Thrush *Ixoreus naevius*
Nashville Warbler *Vermivora ruficapilla*
Townsend's Warbler *Dendroica townsendi*
Connecticut Warbler *Oporornis agilis*
MacGillivray's Warbler *Oporornis tolmiei*
Scarlet Tanager *Piranga olivacea* (s)
Lark Sparrow *Chondestes grammacus* (s)
Golden-crowned Sparrow *Zonotrichia atricapilla* (s)
Pine Grosbeak *Pinicola enucleator*
Red Crossbill *Loxia curvirostra*
White-winged Crossbill *Loxia leucoptera* (s)
Hoary Redpoll *Carduelis hornemanni*

CHECKLIST OF MAMMALS

Names and species sequence are those of *The Mammals of Canada* (A.W.F. Banfield, 1974) with revisions by Hugh C. Smith, former Curator of Mammalogy at the Provincial Museum of Alberta.

This checklist pertains to all of the lake shore as well as surrounding pastures, fields and woods, to the nearest county roads. The 32 species of mammals recorded in the area between 1965 and 1990 have been marked with three different symbols. Species that the visitor can expect to see or that are considered common are checked with a dash (-); species that have often been reported, but on an irregular and unpredictable basis, are indicated by open circles (○); species recorded less than half a dozen times are marked with a solid symbol (●). The four hypothetical species have been reported in the vicinity of the lake, but there are no reliable records for the checklist area.

- Masked Shrew *Sorex cinereus*
○ Arctic Shrew *Sorex arcticus*
○ Little Brown Bat *Myotis lucifugus*
● Hoary Bat *Lasiurus cinereus*
○ Big Brown Bat *Eptesicus fuscus*
- Snowshoe Hare *Lepus americanus*
- White-tailed Jack Rabbit *Lepus townsendii*
- Richardson's Ground Squirrel *Spermophilus richardsonii*
○ Thirteen-lined Ground Squirrel *Spermophilus tridecemlineatus*
● Franklin's Ground Squirrel *Spermophilus franklinii*
○ Northern Flying Squirrel *Glaucomys sabrinus*
- Northern Pocket Gopher *Thomomys talpoides*
- American Beaver *Castor canadensis*
○ Deer Mouse *Peromyscus maniculatus*
○ Southern Red-backed Vole *Clethrionomys gapperi*
- Muskrat *Ondatra zibethicus*
- Meadow Vole *Microtus pennsylvanicus*
○ House Mouse *Mus musculus*

○ Jumping Mouse *Zapua* sp.
○ American Porcupine *Erethizon dorsatum*
- Coyote *Canis latrans*
○ Red Fox *Vulpes vulpes*
○ Ermine *Mustela erminea*
○ Least Weasel *Mustela nivalis*
○ Long-tailed Weasel *Mustela frenata*
○ American Mink *Mustela vison*
○ American Badger *Taxidea taxus*
- Striped Skunk *Mephitis mephitis*
● Canada Lynx *Lynx canadensis*
○ Mule Deer *Odocoileus hemionus*
- White-tailed Deer *Odocoileus virginianus*
● Moose *Alces alces*

Hypothetical:
? Woodchuck *Marmota monax*
? American Red Squirrel *Tamiasciurus hundsonicus*
? American Black Bear *Ursus americanus*
? Cougar *Felis concolor*

CHECKLIST OF FISH, AMPHIBIANS AND REPTILES

Reviewed by Wayne Roberts, Collection Manager of
the Museum of Zoology, University of Alberta

Fathead Minnow *Pimephales promelas*
Brook Stickleback *Culaea inconstans*
Tiger Salamander *Ambystoma tigrinum*
Canadian Toad *Bufo americanus hemiophrys*
Boreal Chorus Frog *Pseudacris triseriata maculata*
Wood Frog *Rana sylvatica*
Common Garter Snake *Thamnophis radix*

TREES, SHRUBS AND WILDFLOWERS MENTIONED IN THE TEXT

Species names based on *Wild Flowers of Alberta*
(R.G.H. Cormack, 1977) and reviewed by Derek
Johnson, Plant Scientist with Forestry Canada,
Edmonton.

White Spruce *Picea glauca*
Balsam Poplar *Populus balsamifera*
Trembling Aspen *Populus tremuloides*
Willow *Salix* sp.
Speckled Alder *Alnus tenuifolia*
White Birch *Betula papyrifera*
Beaked hazelnut *Corylus cornuta*
Manitoba Maple *Acer negundo*

Sago Pondweed *Potamogeton pectinatus*
Giant Reed Grass *Phragmites australis*
Common Cattail *Typha latifolia*
Foxtail Barley *Hordeum jubatum*
Common Great Bulrush *Scirpus validus*
Prairie Crocus *Anemone patens*
Penny Cress, Stinkweed *Thlaspi arvense*

Saskatoon Berry *Amelanchier alnifolia*
Wild Strawberry *Fragaria glauca*
Silverweed *Potentilla anserina*
Choke Cherry *Prunus virginiana*
Prickly Rose *Rosa acicularis*
Wild Red Raspberry *Rubus strigosus*
White Clover *Trifolium repens*
Buffalo Bean *Thermopsis rhombifolia*
Wild Mint *Mentha arvensis*
Snowberry *Symphoricarpos albus*
Buckbrush *Symphoricarpos occidentalis*
Common Yarrow *Achillea millefolium*
Low Everlasting, Pussytoes *Antennaria nitida*
Pasture Sage *Artemisia frigida*
Common Blue Lettuce *Lactuca pulchella*
Gumweed *Grindelia squarrosa*
Marsh Ragwort *Senecio congestus*
Tall Smooth Goldenrod *Solidago gigantea*
Perennial Sow Thistle *Sonchus arvensis*
Common Dandelion *Taraxacum officinale*

REFERENCES

Anderson, R.S. 1970 Shrimps, flees, and soda water. *Alberta Lands, Forests, Fish, and Wildlife Magazine* 13:8-13.

Brechtel, S. 1977. Avian species list for Beaverhill Lake and surrounding uplands. *Alberta Naturalist* 5:168-171.

Dekker, D. 1968. Autumn records of Parasitic Jaegers in central Alberta. *Blue Jay* 26:16-17.

———. 1972. Fall migration of Ravens at Beaverhills Lake. *Edmonton Naturalist* 1(2):7.

———. 1975. Egret at Beaverhills. *Edmonton Naturalist* 3(5):6.

———. 1975. Red Foxes in central Alberta. *Edmonton Naturalist* 3(5):7-8.

———. 1975. Snowy Plover and other uncommon shorebirds at Beaverhills Lake. *Edmonton Naturalist* 3(8):5-6.

———. 1975. Some observations on fall migration of geese at Beaverhills Lake; 1964-1975. *Edmonton Naturalist* 3(10):9-10.

———. 1975. Fluctuations in occurrence and nesting of shorebirds, gulls, geese, herons, pelicans, and cormorants at Beaverhills Lake as a possible result of recent high water levels. *Edmonton Naturalist* 3(10):11-14.

———. 1976. To graze or not to graze the pastures around Beaverhills Lake. *Edmonton Naturalist* 4(5):105-107.

———. 1976. Fish-catching Muskrat. *Blue Jay* 34:151.

———. 1976. Mortality rates of Red Fox pups, and causes of death of adult foxes in central Alberta. *Alberta Naturalist* 6:65-67.

———. 1976. First sight record of Western Sandpiper at Beaverhills Lake, with a note on the field marks of the Least Sandpiper. *Alberta Naturalist* 6:73-75.

———. 1976. Muskrat, Marbled Godwit, and Willet feeding on Sticklebacks at Beaverhills Lake. *Alberta Naturalist* 6:184-185.

———. 1977. Field-identification of Peregrines, Prairie Falcons, and Gyrs in south and central Alberta. *Alberta Naturalist* 7:1-5.

———. 1977. Avocets and habitat. *Edmonton Naturalist* 5(6):144.

———. 1977. Chestnut-collared Longspurs at Beaverhills Lake. *Edmonton Naturalist* 5(6):148.

———. 1977. Smith's Longspurs at Beaverhills Lake. *Edmonton Naturalist* 5(6): 149.

———. 1977. Beaverhills Lake. *Alberta Naturalist* 5:152-153.

———. 1979 Notes from Beaverhills. *Edmonton Naturalist* 7, fall issue.

———. 1979. Long-tailed Jaeger preys on Lesser Yellowlegs. *Blue Jay* 37: 221-222.

———. 1979. Characteristics of Peregrine Falcons migrating through central Alberta, 1969-1978. *Canadian Field-Naturalist* 93:296-302.

———. 1980 Beaverhills Lake. *A Nature Guide to Alberta*. Alberta Provincial Museum Publication No. 5, pp. 183-186.

———. 1980. Hunting success rates, foraging habits, and prey selection of Peregrine Falcons migrating through central Alberta. *Canadian Field-Naturalist* 94:371-382.

———. 1982. Occurrence and foraging habits of Prairie Falcons at Beaverhills Lake, Alberta. *Canadian Field-Naturalist* 96: 477-478.

———. 1982. An introduction to Beaverhills Lake. *Alberta Naturalist* 12:1-5.

———. 1982. This lake is for the birds. *Nature Canada Magazine* 11(3): 17-21.

———. 1983. Denning and foraging habits of Red Foxes, and their interaction with Coyotes in central Alberta, 1972-1981. *Canadian Field-Naturalist* 97:303-306.

———. 1983. Gyrfalcon sightings at Beaverhills Lake and Edmonton, 1964-1983. *Alberta Naturalist* 13:103.

————. 1983. The Bald Eagle - Hunter or scavenger? *Alberta Naturalist* 13:43-45.

————. 1984. Spring and fall migration of Peregrine Falcons in central Alberta, 1979-1983, with comparisons to 1969-1978. *Journal of Raptor Research* 18:92-97.

————. 1984. Golden Eagles at Beaverhills Lake. 1964-1983. *Alberta Naturalist* 14:54-55.

————. 1984. Ancient Murrelet at Beaverhills. *Alberta Naturalist* 14:98.

————. 1984. Migrations and foraging habits of Bald Eagles in east-central Alberta, 1964-1983. *Blue Jay* 42:199-205.

————. 1985. *Wild Hunters*. Canadian Wolf Defenders publication.

————. 1985. Jaegers at Beaverhills Lake. *Alberta Naturalist* 15:1-4.

————. 1987. Veterans of a thousand hunts - Merlins. *Nature Canada Magazine* 16(1):9.

————. 1987. Peregrine Falcon predation on ducks in Alberta and British Columbia. *Journal of Wildlife Management* 51:156-159.

————. 1987. International wetland status for Beaverhills Lake. *Alberta Naturalist* 17:173.

————. 1988. Peregrine Falcon and Merlin predation on small shorebirds and passerines in Alberta. *Canadian Journal of Zoology* 66:925-928.

Dunn, P. 1988. Tree Swallow Research at the BBO. *Beaverhill Bird Observatory Newsletter* 2(1).

Godfrey, W.E. 1986. *The Birds of Canada*. Revised edition. National Museum of Natural Sciences. Ottawa.

Jones, E.T. 1959. Shorebird paradise - Beaverhills Lake. *Blue Jay* 17:100-101.

Jones, R. 1988. Grassland Sparrow Survey. *Beaverhill Bird Observatory Newsletter* 2(1).

Jungkind, S. 1986, 1987. *Annual Reports for the BBO*.

Lister, R. 1979. *The Birds and Birders of Beaverhills Lake*. Edmonton Bird Club Publication.

McNicholl, M. 1977. McCowan's Longspur at Beaverhill Lake. *Alberta Naturalist* 5:147.

Moyles, D. 1987. The Greater Prairie Chicken in Alberta. *Endangered Species in the Prairie Provinces*. G.L. Holroyd et al., eds. Natural History Occasional Paper No. 9. Provincial Museum of Alberta, pp. 217-220.

Murkin, A.R. and D.A. Wrubleski. 1987. Aquatic invertebrates of freshwater wetlands: Function and ecology. *The Ecology and management of wetlands, Volume 1*. D.D. Hook et al, eds. Croom Helm, London.

Nieman, D.J., G.S. Hochbaum, F.D. Caswell, and B.C. Turner. *Monitoring hunter performance in Prairie Canada*. North American wildlife and natural resources conference. Report 52:233-245.

Nyland, E. 1969. This dying watershed. *Alberta Lands, Forests, Parks, and Wildlife Magazine* 12:22-38.

————. 1970. Miquelon Lake. *Alberta Lands, Forests, Parks, and Wildlife Magazine* 13:18-25.

Philips, G.A., ed. 1969. *Tales of Tofield*. Tofield Historical Society publication.

Salt, W.R. and J.R. Salt. 1976. *The Birds of Alberta*. Hurtig Publishers, Edmonton.

Smith, A. 1977. The Sharp-tailed Sparrow at Beaverhill Lake. *Alberta Naturalist* 5:146.

Stepney, P.H.R. 1987. Management considerations for the American White Pelican in Alberta. *Endangered Species in the Prairie Provinces*. G.L. Holroyd et al. eds. Natural History Occasional Paper No. 9. Provincial Museum of Alberta, pp. 155-171.

Weseloh, D.V. 1977. Summer records of Snow Geese at Beaverhill Lake. *Alberta Naturalist* 5:158-159.

Weseloh, D.V., D. Dekker, S. Brechtel, and R. Burns. 1975. Notes on the Double-crested Cormorants, White Pelicans, and Great Blue Herons of Beaverhill Lake, summer 1975. *Alberta Naturalist* 5:132-137.

Weseloh, D.V., S. Brechtel, L. Bogaert, R. Burns, and J. Kreizer. 1977. A survey of the colonial nesting waterbirds at Beaverhill Lake, 1976. *Alberta Naturalist* 5:183-192.

Wrubleski, D.A. 1987. Chironomidae (diptera) of peatlands and marshes in Canada. *Memoirs of the Entomological Society of Canada* 140:141-161.

For more information and field trips to the lake, contact:

Beaverhill Bird Observatory
Box 1418
Edmonton, Alberta
T5J 2N5

Beaverhill Lake Nature Centre
Box 30
Tofield, Alberta
T0B 4J0

Canadian Nature Federation
#606, 1 Nicholas Street
Ottawa, Ontario
K1N 7B7

Canadian Wildlife Service,
Environmental Conservation Branch
Prairie and Northern Region
#200, 4999 - 98 Ave.
Edmonton, Alberta
T6B 2X3

Ducks Unlimited Canada
4912 - 52 Ave.
Tofield, Alberta
T0B 4J0

Edmonton Bird Club
Box 1111
Edmonton, Alberta
T5J 2M1

Edmonton Natural History Club
Box 1582
Edmonton, Alberta
T5J 2N9

Environmental Protection,
Fish and Wildlife Services
10th floor, South Petroleum Plaza,
9915 - 108 St.
Edmonton, Alberta
T5K 2G8

Federation of Alberta Naturalists
Box 1472
Edmonton, Alberta
T5J 2N5

John Janzen Nature Centre
Box 2359
Edmonton, Alberta
T5J 2R7

The Wildbird General Store
4712 - 99 St.
Edmonton, Alberta
T6E 5H5

World Wildlife Fund Canada
90 Eglinton Ave. East, #504
Toronto, Ontario
M4P 2Z7